Lecture Notes in Mathematics

A collection of informal reports and seminars
Edited by A. Dold, Heidelberg and B. Eckmann, Zürich

230

Lucien Waelbroeck

Vrije Universiteit Brussel, Brussel/België
presently at the
Université Libre de Bruxelles, Bruxelles/Belgique

Topological Vector Spaces and Algebras

Springer-Verlag
Berlin · Heidelberg · New York 1971

AMS Subject Classifications (1970): Primary: 46–02, 46 A 15, 46 A 99, 46 H 05, 46 J 05
Secondary: 46 E 40, 46 F 05

ISBN 3-540-05650-5 Springer-Verlag Berlin · Heidelberg · New York
ISBN 0-387-05650-5 Springer-Verlag New York · Heidelberg · Berlin

Offsetdruck: Julius Beltz, Hemsbach/Bergstr.

to Christine

INTRODUCTION

The lectures associated with these notes were given at the Instituto de Matemática Pura e Aplicada (IMPA) in Rio de Janeiro, during the local winter 1970. To emphasize the properties of topological algebras, the author had started out his lectures with results about topological algebras, and introduced the linear results as he went along.

The present exposition is more systematic. It is more or less divided into two parts. The first part contains the first four chapters and could be headed Linear Theory. The second part contains the five remaining chapters and would be headed Algebras.

The author has not intended to write a complete survey of the subject.

Locally convex spaces have become classical. Nuclear spaces are not used in Part II. Both theories have been omitted. Part I mainly contains structures which are not classical, but which are useful in the development of topological algebra theory according to the author's experience. Other mathematicians will feel that this author has been unfair to their own pet theories ...

A limited number of interesting constructions applicable to topological algebras are developed in Part II. It is hoped that the reader will get an idea of the tools with which one can tackle problems involving topological algebras, also of the problems which can be handled in this way. But the reader must be aware that many interesting results will not be found in these notes.

Most chapters are concluded by a section "Notes and Remarks", in which the author speaks about the history, the development of the subject. These Notes are not 100 % reliable. But the author belongs to the old-fashioned school, he believes that the history of a subject is part of the context in which the subject should be placed. An unreliable historical survey is better than none. The amount of work that goes into the writing of a reliable historical survey is incompatible with the "Lecture Note" idea. This does not mean that the author should not try to help the mathematician who later wishes to write a reliable historical survey.

All Lecture Note authors have to meet deadlines, and decide to leave out material which definitely should be included. This is a heart-rending experience.

N.Adasch [1], [2], [3] has proved nice results about non locally convex, ultrabarreled spaces, and about the closed graph theorem "à la Ptak". These results would fit very well in the first chapter.

The author would also have included some personal results about the density, the admissibility problem of V.Klee [44], [45]. The problem has also been studied by A.H.Shuchat [70]. The author owes references to R.Retherford, these will be very useful to him when he has the leisure to write out his results.

The author must acknowledge support from the Organization of American States during his stay in Rio de Janeiro. His stay at IMPA was quite agreable. The friendly astmosphere, the working conditions were conducive to scientific work. The scheduling of the lectures was leisurely enough, but not too leisurely.

These lecture notes have been typed by the secretaries of the Departments of Mathematics of the Université Libre de Bruxelles and the Vrije Universiteit Brussel. The author is especially grateful to the U.L.B. and its mathematics department secretaries, who did a large part of the job, since he does not belong to that institution.

Dr Ivan Cnop corrected a first draft of these notes. His help was greatly appreciated.

TABLE OF CONTENTS

CHAPTER I

TOPOLOGICAL VECTOR SPACES

1. Topological vector spaces : Let E be a real or a complex vector space : the scalar field will be called $\mathbb{K}$ throughout this paragraph. Let $\mathcal{T}$ be a topology on E.

Definition 1 : $\mathcal{T}$ is a vector space topology if the mappings $(x, y) \to y - x$, $E \times E \to E$ and $(s, x) \to s \,.\, x$, $\mathbb{K} \times E \to E$ are continuous.

A vector space topology is of course determined by the filter $\mathcal{V}$ of neighbourhoods of the origin. The following is well known, its proof will not be given here.

Proposition 1 : $\mathcal{V}$ is the filter of neighbourhoods of the origin for some vector space topology on E iff it has the following properties :
a) $\mathcal{V}$ has a balanced basis
b) the elements of $\mathcal{V}$ are absorbing
c) for every $V \in \mathcal{V}$ some $V' \in \mathcal{V}$ can be found such that $V \supseteq V' + V'$.

The reader may be reminded that $A \subseteq E$ is balanced when $s\,A \subseteq A$ for every $s \in \mathbb{K}$, $|s| \leqslant 1$. Also, $A \subseteq E$ is absorbing if we can associate to every $x \in E$ some $\varepsilon > 0$ such that $s\,x \in A$ when $|s| \leqslant \varepsilon$. The set of absorbing subsets of E is a filter with balanced basis.

Definition 2 : Let E be a real or a complex vector space. An $\mathcal{J}$-seminorm on E is a mapping $\nu : E \to \mathbb{R}_+$ such that :
a) $\nu(sx) \leqslant \nu(x)$ when s is scalar, $|s| \leqslant 1$

b) $\nu(sx) \to 0$ when s is scalar, $s \to 0$ and $x \in E$ is constant

c) $\nu(x + y) \leq \nu(x) + \nu(y)$

If furthermore $\nu(x) \neq 0$ when $x \neq 0$, ν is an $\mathfrak{F}$-norm.

A vector space topology can be determined by a family of $\mathfrak{F}$-semi-norms. We obtain a fundamental system of neighbourhoods of the origin in this topology by selecting a finite number of indices $i_1,\ldots,i_k$ and some $\varepsilon > 0$, and letting.

$$V_{i_1 \ldots i_k} = \{x \in E \mid \nu_{i_j}(x) \leq \varepsilon\} \quad \text{for} \quad j = i,\ldots,k\}$$

A countable family ν_n, $n \in \mathbb{N}$ of $\mathfrak{F}$-semi-norms can always be replaced by a single one :

$$\nu(x) = \Sigma\, 2^{-n} \frac{\nu_n(x)}{1 + \nu_n(x)}$$

for instance.

<u>Proposition 2</u> : A vector space topology on E can always be determined by a family of $\mathfrak{F}$-semi-norms. If the topology is metrizable, then it can be determined by a single $\mathfrak{F}$-norm.

Let V be any balanced neighbourhood of the origin in E. Define $V_1,\ldots, V_k,\ldots$ inductively in such a way that each V_k is balanced, $V \supseteq V_1 + V_1$, and $V_k \supseteq V_{k+1} + V_{k+1}$. Let $\alpha = q/2^N$ be a dyadic rational, $0 \leq \alpha < 1$, then $\alpha = \Sigma_1^N t_k\, 2^{-k}$ with $t_k = 0$ or 1 for all k. Put

$$W_\alpha = \Sigma_1^N t_k V_k$$

i.e.

$$W_\alpha = \Sigma_{t_k = 1} V_k$$

Then W_α is balanced, and clearly $W_{\alpha_1} + W_{\alpha_2} \subseteq W_{\alpha_1+\alpha_2}$ if $\alpha_1 + \alpha_2 < 1$. We put $W_\alpha = E$ for $\alpha \geq 1$, and define for $x \in E$

$$q(x) = \inf \{\alpha \mid x \in W_\alpha\}$$

Then q is an $\mathfrak{J}$-semi-norm.

If the space E is metrizable, we choose the neighbourhoods V_k in such a way that they constitute a fundamental system of neighbourhoods of the origin. The constuction above gives an $\mathfrak{J}$-semi-norm which determines the topology. Since the topology is Hausdorff the $\mathfrak{J}$-semi-norm is an $\mathfrak{J}$-norm.

Definition 3 : A subset B of a topological vector space is bounded if we can associate to every neighbourhood U of the origin some $\varepsilon > 0$ such that $\varepsilon B \subseteq U$.

2. Convex and pseudo-convex topologies : Let $0 < p \leq 1$

Definition 4 : A p-semi-norm on a real or a complex vector space E is a mapping $\nu : E \to \mathbb{R}_+$ such that

a. $\nu(\lambda x) = |\lambda|^p \nu(x)$ for λ scalar, $x \in E$.

b. $\nu(x + y) \leq \nu(x) + \nu(y)$

If $\nu(x) \neq 0$ for $x \neq 0$, ν is a p-norm.

Definition 5 : A subset A of the vector space E is absolutely p-convex if $\lambda x + \mu y \in A$ when $x \in A$, $y \in A$, $|\lambda|^p + |\mu|^p \leq 1$
The absolutely p-convex hull of $A_o \subseteq E$ is the set

$$co_p A_o = \{\Sigma_1^N \lambda_n x_n | \forall n : x_n \in A_o, \Sigma |\lambda_n|^p \leq 1\}$$

Definition 6 : The p-homogeneous Minkowski functional of an absolutely p-convex set A is function

$$\nu_A(x) = \inf \{|\lambda|^p \mid x \in \lambda A\}$$

It is clear that the absolutely p-convex hull of A_o is absolutely p-convex, and is the smallest absolutely p-convex set containing A_o. The p-homogeneous Minkowski functional of an absolutely p-convex set is a p-semi-norm. The proof is standard when $p = 1$ the standard proof requires hardly any change for other values of p. Select λ, μ in such a way that $|\lambda|^p > \nu_A(x)$, $|\mu|^p > \nu_A(y)$
then $x \in \lambda A$, $y \in \mu A$
Then

$$\frac{x + y}{(\lambda + \mu)^p} = \frac{\lambda^p}{(\lambda + \mu)^p} \frac{x}{\lambda^p} + \frac{\mu^p}{(\lambda + \mu)^p} \frac{y}{\mu^p}$$

is a p-convex combination of elements of A. It belongs therefore to A and $\lambda + \mu \geqslant \nu_A(x + y)$

We shall call a set absolutely pseudo-convex if it is absolutely p-convex for some $p > 0$. Similarly ν will be a pseudo-semi-norm if it is a p-semi-norm for some p.

Definition 7 : A vector space topology is locally p-convex if the origin has a fundamental set of absolutely p-convex neighbourhoods. It is locally pseudo-convex if the origin has a fundamental system of absolutely pseudo-convex neighbourhoods.

We see that a locally p-convex topology can be determined by p-semi-norms, and that a locally pseudo-convex topology can be defined by pseudo-semi-norms.

Proposition 3 : Let E be a real or a complex vector space. Let A be a balanced subset of E . Assume that $A + A \subseteq 2^{1/p} A$ for some $p > 0$.. Then

$$\mathrm{co}_p\, A \subseteq 2^{1/p} A$$

We consider some $x \in \mathrm{co}_p\, A$, then $x = \Sigma_1^N \lambda_n a_n$ with $\Sigma_n |\lambda_n|^p \leqslant 1$, and with all $a_n \in A$. We choose k_n in such a way that

$2^{-k_n} \geqslant |\lambda_n|^p > 2^{-k_n - 1}$. Then $\Sigma\, 2^{-k_n} \leqslant 2$.

We now define $a'_n = 2^{-k_n/p} \lambda_n a_n$ for $n = 1, \ldots, N$, and add a few terms $a'_n = 0$, for $n = N + 1, \ldots, N'$ and corresponding intergers k_n so that $\Sigma_1^{N'} 2^{-k_n} = 2$. Then

$$x = \Sigma_1^{N'} 2^{-k_n/p} a'_n$$

Once this has been done, we may reason by induction. We reorder the terms so that $k_1 \geqslant k_2 \geqslant \ldots. \geqslant k_{N'}$. Then $k_1 = k_2$ unless $N' = 1$, because $\Sigma 2^{-k_n} = 2$, and there is nothing to prove when $N' = 1$. We let

$$b = 2^{-1/p} a'_1 + 2^{-1/p} a'_2$$

so $b \in A$, and

$$x = 2^{(-k_1+1)/p} b + \Sigma_3^{N'} 2^{-k_n/p} a'_n$$

This is an expression for x, which is completely similar to the one we had previously, but with only $N' - 1$ terms.

Corollary 1 : A locally bounded Hausdorff topological vector space is p-normable for some $p > 0$.

Let B be a bounded balanced neighbourhood of the origin. Then $B + B$ is bounded, so $B + B \subseteq MB$ for some M. It is clear that $M \geqslant 2$, unless B is a vector space, so we may put $M = 2^{1/p}$ and $0 < p \leqslant 1$ ($p = \log 2/\log M$). The set $co_p B$ is absolutely p-convex; its Minkowski functional is a p-norm which defines the topology of the space.

<u>Corollary 2</u> : A topological vector space is locally pseudo-convex if we can find a fundamental system V_i of neighbouhodds of the origin, and for each V_i some M_i in such a way that $V_i + V_i \subseteq M_i V_i$.

Just define p_i so $2^{1/p_i} = M_i$, and let $W_i = co_{p_i} V_i$. The sets W_i and their multiples are a fundamental set of neighbourhoods of the origin, and are pseudo-convex.

3. <u>Direct limits</u> : Let E be a real or a complex vector space. Let $\Phi_1, \Phi_2, \ldots$ be a decreasing sequence of filters with balanced bases on E. Assume that $\Sigma'_n F_n = \cup_n \Sigma_1^n F_k$ is an absorbing subset of E whenever $F_n \in \Phi_n$ for all n.

<u>Proposition 4</u> : The sets $\Sigma'_n F_n$ are the basis of neighbouhoods of the origin in the strongest vector space topology $\mathcal{T}$ on E such that $\Phi_n \to 0$ for all n.

These sets are the basis of a balanced, absorbing filter $\mathcal{U}$ which is weaker than each Φ_n. If $U \in \mathcal{U}$, then $U \in \Sigma' F_n$ where each $F_n \in \Phi_n$. Let $G_n = F_{2n-1} \cap F_{2n}$, then $G_n \in \Phi_n$. $V = \Sigma' G_n \in \mathcal{U}$, and $U \supseteq V + V$. This shows already that $\mathcal{U}$ is the filter of neighbourhoods of the origin in a vector space topology $\mathcal{T}$ such that each $\Phi_n \to 0$.

Let now $\mathcal{T}'$ be another vector space topology on E, such that $\Phi_n \to 0$ for all n, let U be a neighbourhood of the origin in $\mathcal{T}'$, then $V_1, V_2, \ldots$ be neighbourhoods such that $U \supseteq V_1 + V_1, \ldots, V_{k-1} \supseteq V_k + V_k, \ldots$ so that $U \supseteq \Sigma' V_n$. Of course $V_n \in \Phi_n$ for all n, so $U \in \mathcal{U}$ and $\mathcal{T}'$ is weaker than $\mathcal{T}$.

Note 1 : If $\Phi_1, \Phi_2, \ldots$ is not a decreasing sequence of filters, we can consider $\Phi_1, \Phi_1 \cap \Phi_2, \ldots$ which is decreasing. If the given filters Φ_n do not have balanced bases, we can replace each Φ_n by the strongest balanced filter Φ'_n weaker than Φ_n; Φ'_n is generated by the balanced hulls of the elements of Φ_n. So neither of these two conditions is really restrictive. They are useful however to simplify the statement of proposition 1.

Note 2 : It is of course necessary that the sets $\Sigma' F_n$ be absorbing if we want them to generate a vector space topology. If the condition is not fullfilled by the given filters Φ_n, then we should replace Φ_n by a weaker filter Φ'_n, such that $\Phi'_n \to 0$ in any vector space topology in which $\Phi_n \to 0$. Letting for instance Ψ be the filter of absorbing subsets of E, we can let $\Phi'_n = \Phi_n \cap \Psi$.

But other choices may be preferable and a little ingeniosity may be useful in practice.

Corollary 1 : Let E be a vector space, let E_n be an increasing sequence of subspaces of E with $E = \bigcup_n E_n$; for each n let $\mathcal{T}_n$ be a vector space topology on E_n, and assume that the identity mapping $E_n \to E_{n+1}$ is continuous. The sets $\Sigma'_n U_n$, where each U_n is a neighbourhood of the origin in $(E_n, \mathcal{T}_n)$, range over a fundamental system of neighbourhoods of the origin in the direct limit vector space topology of E (i.e. the strongest vector space topology on E which induces on each E_n a weaker topology than $\mathcal{T}_n$).

Corollary 2 : Assume that E_n is locally convex, or locally p-convex for all n. Under the assumptions of corollary 1, the direct limit vector space topology of E is locally convex, or locally p-convex respectively.

Corollary 1 is a simple application of proposition 1, we want Φ_n to tend to zero for all n, where Φ_n is the filter of neighbourhoods fo the origin in E_n. Corollary 2 is obvious, since $\Sigma' U_n$ is absolutely convex, or absolutely p-convex when each U_n is absolutely convex or absolutely p-convex.

<u>Corollary 3</u> : Let E be a vector space, B_n an increasing sequence of balanced subsets of E. Assume that $\cup_n B_n$ generates E. The sets $\Sigma'_n \varepsilon_n B_n$, where ε_n ranges over sequences of strictly positive real numbers, generate the filter of neighbourhoods of the origin in the strongest vector space topology for which all sets B_n are bounded.

B_n is bounded in a vector space topology if $s B_n \to 0$ when s is scalar, $s \to 0$. We can therefore apply proposition 1 to determine the topology of E.

We must still check that the sets $\Sigma'_n \varepsilon_n B_n$ are absorbing. Let $x \in E$, then $x = \Sigma_1^N \lambda_k b_k$ where each $b_k \in \cup_n B_n$. For each k, we choose n_k such that $b_k \in B_{n_k}$; since the sequence of sets B_n is increasing, we can assume that the mapping $k \to n_k$ is injective. Then $\alpha x \in \Sigma'_n \varepsilon_n B_n$ where $\alpha = \min (\varepsilon_{n_k} / \lambda_k)$.

<u>Note 3</u> : Proposition 1 can also be applied to determine the strongest vector space topology on E in which an uncountable family of filters Φ_i converge to zero. It is the strongest vector space topology in which $\cap_i \Phi_i = \Phi$ converges to zero. The filter Φ has the elements $\cup_i F_i$ where each $F_i \in \Phi_i$.

Corollary 4 : Let E be a vector space, for each $i \in I$, let E_i be a subspace of E, assume $\cup E_i$ generates E. Let $\mathcal{T}_i$ be a vector space topology on E_i. The sets

$$\Sigma'_n \cup_i V_{i_n}$$

where V_{i_n} is a neighbourhood of the origin in E_i, for each i and each n, generate the filter of neighbourhoods of the origin of the strongest vector space topology which induces on each E_i a weaker topology than $\mathcal{T}_i$.

Corollary 5 : Let E be a vector space, $\mathcal{B}$ a family of balanced subsets of E, $\cup_{B\in\mathcal{B}} B$ generating E. The sets

$$\Sigma'_n \cup_{B\in\mathcal{B}} \varepsilon(n, B).\ B$$

range over a fundamental system of neighbourhoods of the origin in the strongest vector space topology on E for which all $B \in \mathcal{B}$ are bounded.

Corollaries 4 and 5 are obvious. Their proof is left to the reader.

Note 4 : An uncountable vector space direct limit topology is usually not locally convex. As a matter of fact, an uncountable direct sum is never a locally convex space. The easiest is to show that the strongest vector space topology on $\mathbb{R}^{(I)}$, a space of uncountable dimension is not locally convex, but the set of families (x_i) where $\Sigma\,|x_i|^p \leq 1$ is a neighbourhood of the origin, and no convex subset of this set meets more than a countable number of coordinate axes off the origin.

Consideration of the $\mathcal{J}$-norm $\Sigma\, \rho(x_i)$, where ρ is positive, subadditive, and such that $t^p/\rho(t) \to 0$ with t for all positive p would show similarly that $\mathbb{R}^{(I)}$ is not locally pseudo-convex.

4. Ultrabarreled spaces : Definition 8

Let $(E,\mathcal{T})$ be a topological vector space. $(E,\mathcal{T})$ is ultrabarreled if a pointwise bounded set $\{u_i\}_{i\in I}$ of continuous linear mappings of E into a topological vector space $(F,\mathcal{S})$ is always equicontinuous.

The standard examples of ultrabarreled spaces are well-known, or easy to construct once the idea of a barreled space has been understood. Baire topological vector spaces are ultrabarreled, so are direct limits of ultrabarreled spaces, in the vector space direct limit topology. Quite a few locally convex spaces are ultrabarreled; all locally convex Fréchet spaces, and all countable direct limits of ultrabarreled locally convex spaces are ultrabarreled and locally convex.

Proposition 5 : A vector space topology $\mathcal{T}$ on E is ultrabarreled iff it is stronger than any vector space topology $\mathcal{T}_1$ on E, in which the origin has a fundamental system of $\mathcal{T}$-closed neighbourhoods.

This proposition can be reworded.

Proposition 5' : $\mathcal{T}$ is ultrabarreled iff the sets V_1, V_2, ... are neighbourhoods of the origin in E as soon as they are closed, balanced, absorbing, and $V_k \supseteq V_{k+1} + V_{k+1}$ for all k.

Let us first show why propositions 5 and 5' are equivalent. A sequence $V_1, \ldots, V_k, \ldots$ of balanced absorbing subsets of E is a fundamental sequence of neighbourhoods of the origin in some vector space topology if $V_k \supseteq V_{k+1} + V_{k+1}$. If these sets are closed, and if the topology satisfies the condition of proposition 5, then this vector space topology is weaker than $\mathcal{T}$ and V_k is a neighbourhood of the origin for all k. Assume conversely that the condition of proposition 5' is satisfied, let V_1 be a neighbourhood of the origin in a topology $\mathcal{T}_1$ satisfying the condition of proposition 5, and define $V_2, \ldots, V_k, \ldots$ in such a way that for all k, $V_k \supseteq V_{k+1} + V_{k+1}$, each V_k being $\mathcal{T}$-closed and balanced. Then $V_1, V_2, \ldots$ are neighbourhoods of the origin and $\mathcal{T}_1$ is weaker than $\mathcal{T}$.

It is nearly clear that either of these conditions implies that $\mathcal{T}$ is an ultrabarreled topology. Let $u_i : (E, \mathcal{T}) \to (F, \mathcal{S})$ be a pointwise bounded family of continuous linear mappings.
For each neighbourhood V of the origin, let

$$\phi(V) = \cap_i \, u_i^{-1} V$$

The sets $\phi(V)$ are then a fundamental set of neighbourhoods of the origin in E for some vector space topology $\mathcal{T}_1$; and the origin has a fundamental system of $\mathcal{T}$-closed neighbourhoods. So $\mathcal{T}_1$ is weaker than $\mathcal{T}$, and $\cap u_i^{-1} V$ is a $\mathcal{T}$-neighbourhood of the origin whenever V is an $\mathcal{S}$-neighbourhood.

We must now show the converse. $(E, \mathcal{T})$ is a topological vector space. $V_1, \ldots, V_k, \ldots$ is a sequence of closed, balanced, absorbing subsets of E. We shall construct a metrizable vector space $(F, \mathcal{S})$, $W_1, \ldots, W_k, \ldots$ being a fundamental sequence of neighbourhoods of the origin, and a pointwise bounded family of continuous mappings $u_\phi : E \to F$ in such a way that

$$V_k = \cap_\phi \, \phi^{-1} W_k$$

This will show that the sets V_k are neighbourhoods of the origin if E is ultrabarreled according to definition 8.

ϕ will range over the set Φ of all sequences $(U_1, \ldots, U_k, \ldots)$ of balanced neighbourhoods of the origin such that $U_k \supseteq U_{k+1} + U_{k+1}$ for all k. For each $\phi \in \Phi$, we consider the sequence $(U_1 + V_1, \ldots, U_k + V_k, \ldots)$, we let $G_\phi = \cap_k (U_k + V_k)$, observe that G_ϕ is a vector subspace of E, and $U_k + V_k$ is a union of cosets of G_ϕ.

F_ϕ will be the quotient E/G_ϕ, W_k will be the quotient image of $U_k + V_k$ in F_ϕ. Further F will be the direct sum $\oplus_{\phi\in\Phi} F_\phi$ and W_k the set of families $(F_\phi)_{\phi\in\Phi}$ which belongs to F, and are such that each F_ϕ belongs to $W_{k.\phi}$ (i.e. for all ϕ, $F_\phi \in W_{k.\phi}$, and for all ϕ except a finite number of exceptions, $F_\phi = 0$). Then (W_k) is the basis of a balanced, absorbing filter on F, $W_k \supseteq W_{k+1} + W_{k+1}$ so this is the filter of neighbourhoods of the origin in some vector space topology on F. This vector space topology is Hausdorff (because we have killed off the intersection of the neighbourhoods of the origin) and metrizable.

Now, $u_\phi : E \to F$ will be the composition of the quotient map $E \to F_\phi$ with the canonical embedding $F_\phi \to F$. We observe that $u_\phi^{-1}(W_k) = U_{k\phi} + V_k$. This shows that each u_ϕ is continuous, because the sets $U_{k\phi}$ are neighbourhoods of the origin, and also, that this family of mappings is strongly bounded because each set V_k is absorbing. If E is ultrabarreled according to definition 8,

$$\cap_\phi \bar{u}_\phi^{-1} W_k = \cap (U_{k\phi} + V_k)$$

is a neighbourhood of the origin. We know that this intersection is equal to V_k.

Proposition 5 can be restated in yet another way.

Proposition 5" : A vector space topology $\mathcal{T}$ on E is ultrabarreled if, and only if , each lower semi-continuous $\mathfrak{J}$-semi-norm on E is continuous.

A lower semi-continuous $\mathfrak{J}$-semi-norm ν on E determines on E a topology $\mathcal{T}_1$ such that the origin has a fundamental system of $\mathcal{T}$-closed neighbourhoods. $\mathcal{T}_1$ is weaker than $\mathcal{T}$ if $\mathcal{T}$ is ultra-barreled, and ν is continuous.

Assume conversely that all lower-semi-continuous $\mathfrak{J}$-semi-norms are continuous. Let $V_1, V_2, \ldots, V_k, \ldots$ be balanced, closed, absorbing subsets of E such that $V_k \supseteq V_{k+1} + V_{k+1}$. When α is a dyadic rational, define W_α as in the proof of proposition 2 and let W'_α be the $\mathcal{T}$-closure of W_α. Consider then

$$p(x) = \inf \{\alpha \mid x \in W'_\alpha\}$$

This is a lower semi-continuous $\mathfrak{J}$-semi-norm, so p is continuous and $V_k = \{x \mid p_k(x) \leq 2^{-k}\}$ is a neighbourhood of the origin.

Proposition 6 : Let E be an ultrabarreled space, and F a Fréchet space. A linear map with closed graph $u : E \to F$ is continuous.

The closures of the sets $\overset{-1}{u} V$, V a neighbourhood of the origin in F, are the basis of the filter of neighbourhoods of the origin for some linear topology on E. Since E is ultrabarreled, this topology is weaker than the original one, i.e. the closure of $\overset{-1}{u} V$ is a neighbourhood of the origin in E whenever V is a neighbourhood of the origin in F. The result follows from the following well known result.

Lemma 1 :An almost continuous linear mapping of a topological vector space into a Fréchet space is continuous if it has a closed graph.

We include a proof for completeness's sake. Let $V_1,\dots,V_k,\dots$ be a fundamental sequence of closed, balanced of neighbourhoods of the origin in F, such that $V_k \supseteq V_{k+1}+V_{k+1}$ for all k, and x be in the closure of $u^{-1}\, V_k$. We will show that x is in $u^{-1}\, V_{k-1}$. Choose a_k in E with $ua_k \in V_k$, $x-a_k$ in the closure of $u^{-1}\, V_{k+1}$, and by induction for $\ell > k$, choose a_ℓ with $ua_\ell \in V$ and $x-a_k-\dots-a_\ell$ in the closure of $u^{-1}\, V_\ell$. This is possible since the closures of the sets $u^{-1}\, V_\ell$ are neighbourhoods of the origin.

The series $\Sigma_k a_\ell$ converges because F is complete. Let y be its sum, $y \in V_{k-1}$ because $\Sigma_k a_\ell$ is a limit of elements in V_k+V_k. We will see that ux = y. Since the graph of u is closed, it is sufficient to show that each neighbourhood of (x,y) meets the graph.

Let U be a neighbourhood of x, and let ℓ be large. Choose x' so that $x-x' \in U$, $x'-a_k-\dots-a_\ell \in u^{-1}V_\ell$. Then (x',ux') is in the graph, x-x' is in U and ux'-y is in $V_\ell+V_\ell$ so (x,y) is in the closure of the graph.

Let now E be a vector space and q an $\mathfrak{J}$-semi-norm on E. We let E_q be the quotient of E by the subspace on which q vanishes, and $\hat{E}_q$ be the completion of E_q, when we consider on E_q the $\mathfrak{J}$-norm induced by q. The extension of this induced $\mathfrak{J}$-semi-norm to $\hat{E}_q$ will be called $\hat{q}$.

<u>Lemma 2</u> : Assume the $\mathfrak{J}$-semi-norm q is lower semi-continuous on a topological vector space $(E, \mathcal{T})$. The canonical mapping $u = E \to \hat{E}_q$ then has closed graph.

We let (x,y) be in the closure of the graph, and choose a directed

system $x_i \to x$, $ux_i \to y$. We also choose $y' \in E_q$ with $\hat{q}(y - y') < \varepsilon$, and $y_1' \in E$ with $uy_1' = y'$.
This is possible since $uE = E_q$ and E_q is dense in $\hat{E}_q$.
We must evaluate $\hat{q}(y - ux)$, but

$$\begin{aligned}
\hat{q}(y - ux) &\leq \hat{q}(y - y') + \hat{q}(y' - ux) \\
&= \hat{q}(y - y') + q(y_1' - x) \\
&\leq \hat{q}(y - y') + \liminf q(y_1' - x_i) \\
&= \hat{q}(y - y') + \liminf \hat{q}(uy_1' - ux_i) \\
&\leq 2\varepsilon
\end{aligned}$$

because $\hat{q}(y - y') \leq \varepsilon$, and $\hat{q}(uy_1' - ux_i) \to \hat{q}(uy_1' - y) = \hat{q}(y' - y)$ so both terms are less than ε.

Proposition 7 : A topological vector space $(E, \mathcal{T})$ is ultrabarreled if, and only if, every linear mapping with closed graph of E into a Fréchet space is continuous.

The condition is necessary by proposition 6. And if the condition is satisfied, then by lemma 2, every lower semi-continuous $\mathfrak{J}$-semi-norm is continuous, i.e the space is ultrabarreled (Proposition 5").

5. Ultrabornological and quasi-ultrabarreled spaces : In locally convex space theory, the concept of a bornological topology is standard. So is that of an infrabarreled space, of a space in which a set of linear forms is equicontinuous when it is uniformly bounded on bounded sets.

If E is a locally convex space, and $\mathfrak{S}$ is any family of bounded subsets of E, one defines the bornological topology associated to

$\mathfrak{S}$ as the strongest locally convex topology for which all elements of $\mathfrak{S}$ are bounded. And E is $\mathfrak{S}$-barreled if a barrel which absorbs all elements of $\mathfrak{S}$ is a neighbourhood of the origin.

We shall define here analogues for these classes of spaces. The proofs will be variations of those given in paragraph 4. They will be left to the reader.

For aesthetic reasons, we follow S.O. Iyahen and call quasi-ultrabarreled the spaces one would logically call infra-ultrabarreled. The ultrabornological spaces considered here are not those that have been considered by Bourbaki[10], i.e. locally convex direct limits of Banach spaces. Bourbaki's ultrabornological spaces had been called strictly bornological by Grothendieck [38]. We revert to the earlier terminology and speak of strictly ultrabornological spaces.

We shall use the following word in this paragraph. If B is an absolutely convex subset of a vector space E, and if the Minkowski functional of B is a Banach space norm, then B is <u>completant</u>. Completant sets will be investigated in more detail further(chapter II)

<u>Definition</u> 9 : A vector space topology $\mathcal{T}$ is ultrabornological if it is stronger than any vector space topology $\mathcal{T}'$ such that all $\mathcal{T}$-bounded sets are $\mathcal{T}'$-bounded. If $\mathcal{T}$ is any vector space topology on E, the ultrabornological topology $\mathcal{T}'$ associated to $\mathcal{T}$ is the upper bound of the vector space topologies on E for which all $\mathcal{T}$-bounded sets are bounded. More generally, if $\mathfrak{S}$ is any family of sets, the ultrabornological topology associated to $\mathfrak{S}$ is the strongest vector space topology for which all elements of $\mathfrak{S}$ are bounded.

<u>Definition 10</u> : A vector space topology $\mathcal{T}$ on E is strictly ultrabornological if $\mathcal{T}$ is the ultrabornological topology associated to the family of completant bounded subsets of $(E, \mathcal{T})$.

It is obvious that a metrizable vector space topology $\mathcal{T}$ is ultrabornological (the proof does not differ from the usual one).
It may be less obvious that such a topology is $\mathfrak{S}$-ultrabornological where $\mathfrak{S}$ is the family of bounded convex subsets, and that a Fréchet topology is strictly ultrabornological. These results follow from the fact that a sequence (x_n) which converges to zero has a subsequence (x_{n_k}) such that $(k\, x_{n_k})$ has a bounded convex hull (see further, chapter II, paragraph 5).

Direct limits of ultrabornological spaces are ultrabornological.

The locally convex metrizable spaces and their countable direct limits are standard examples of locally convex ultrabornological spaces. Fréchet locally convex spaces, and countable direct limits of Fréchet locally convex spaces are locally convex, and strictly ultrabornological.

<u>Proposition 8</u> : The following statements are equivalent.
(i) $\mathcal{T}$ is the ultrabornological topology associated to the family $\mathfrak{S}$ of subsets of the vector space E.

(ii) The $\mathcal{T}$-neighbourhoods of the origin absorb all $S \in \mathfrak{S}$; the set V_1 is a neighbourhood of the origin if it is possible to find balanced, abosorbing sets $V_2, \ldots, V_k \ldots$ such that each V_k absorbs all elements of $\mathfrak{S}$ and $V_k \supseteq V_{k+1} + V_{k+1}$ for $k = 1, 2, \ldots$

(iii) The $\mathfrak{J}$-semi-norm ν on E is continuous iff for all $S \in \mathfrak{S}$

$$\sup \{\nu(z\, S) \mid s \in S\} \to 0$$

when $z \to 0$.

(iv) A linear mapping $u : E \to F$ of E into a topological vector space F is continuous when it maps each $S \in \mathfrak{S}$ onto a bounded subset of F.

If $\mathcal{T}$ is the ultrabornological topology associated to $\mathfrak{S}$, a family $u_\alpha : E \to F$ of continuous linear mappings is equicontinuous when $\cup_\alpha u_\alpha S$ is a bounded subset of F for all elements of E.

These results are trivial. They are stated for reference purposes.

<u>Definition 11</u> : Let $\mathfrak{S}$ be a family of bounded subsets of $(E, \mathcal{T})$ where $\mathcal{T}$ is a vector space topology on E, and assume that $E = \cup_{s \in \mathfrak{S}} S$. The topology $\mathcal{T}$ is $\mathfrak{S}$-ultrabarreled if a family u_i of continuous linear mappings of F into a topological vector space F is equicontinuous when it is equibounded on $\mathfrak{S}$. Also $\mathcal{T}$ is quasi-ultrabarreled if it is $\mathfrak{S}$-ultrabarreled where $\mathfrak{S}$ is the family of all bounded subsets of $(E, \mathcal{T})$.

<u>Proposition 9</u> : Assume that $\mathfrak{S}_2 \supseteq \mathfrak{S} \supseteq \mathfrak{S}_1$ where the elements of $\mathfrak{S}_1$ are completant, and $\mathfrak{S}_2$ is the set of all bounded subsets in the strict-ultrabornological topology associated to $\mathfrak{S}_1$. A vector space topology $\mathcal{T}$ is then ultrabarreled when it is $\mathfrak{S}$-ultrabarreled.

<u>Proposition 10</u> : The ultrabornological topology associated to a family $\mathfrak{S}$ of bounded sets is $\mathfrak{S}$-ultrabarreled.

Proposition 10 gives us many examples of $\mathfrak{S}$-ultrabarreled spaces. Proposition 9 allows us to conclude that such an $\mathfrak{S}$-ultrabarreled space is often ultrabarreled. Note that the following criterion follows from proposition 9.

Corollary : Let E be an union of Fréchet spaces E_i, and $\mathfrak{S}$ be the set of subset of E which are contained in one of the E_i and bounded for the Fréchet topology of E_i. A vector space topology $\mathcal{T}$ on E, which induces on each E_i a weaker topology then the given one, is ultrabarreled if it is $\mathfrak{S}$-ultrabarreled.
$\mathfrak{S}_1$ would be the set of completant elements of $\mathfrak{S}$.

We have conditions for $\mathfrak{S}$-ultrabarreledness, analogous to the conditions for ultrabarreledness or for ultrabornologicality.

Proposition 11 : The following conditions are equivalent, where $\mathcal{T}$ is a vector space topology on E, and $\mathfrak{S}$ a family of bounded subsets of E covering E :
(i) $\mathcal{T}$ is $\mathfrak{S}$-ultrabarreled.

(ii) $V_1, V_2, \ldots$ are neighbourhoods of the origin when each V_k is a closed balanced set which absorbs all elements of $\mathfrak{S}$ and $V_k \supseteq V_{k+1} + V_{k+1}$ for all k.

(iii) A lower semi-continuous $\mathfrak{F}$-semi-norm ν is continuous when

$$\sup_{s \in S} \nu(z\, s) \to 0$$

for $z \to 0$, this for all $S \in \mathfrak{S}$

(iv) A linear mapping with closed graph of E into a Fréchet space F is continuous if it maps each element of $\mathfrak{S}$ onto a bounded set.

The proofs of these results are standard variants of the proofs given in paragraph 4. They are left to the reader.

Note 1 : The results in this paragrap, about σ-ultrabarreled spaces, are generalisations of those given in paragraph 4, since an σ-ultrabarreled space is just ultrabarreled when σ is the collection of all finite subsets of E.

Note 2 : It is well known, in locally convex space theory, that a complete quasi-barreled space is barreled. Weaker completeness conditions are sufficient. The proof of this theorem relies strongly on the fact that a locally convex space has many convex bounded sets.

Proposition 9 looks like the natural generalization of this result.

The following example of a quasi-ultrabarreled, complete, non ultrabarreled space is due to P.Turpin.

$\mathbb{R}^{(\mathbb{N})}$ is the direct sum of a countable number of copies of $\mathbb{R}$.
Let $u_k \in \mathbb{R}^{(\mathbb{N})}$ be the"basis vector of order k", i.e. the components $u_{k.i}$ of u_k vanish for all $i \in \mathbb{N}$, $i \neq k$, but $u_{k.k} = 1$.
Let $S = \{u_k \mid k \in \mathbb{N}\}$ and consider on $\mathbb{R}^{(\mathbb{N})}$ the strongest vector space topology for which S is bounded. This topology is ultrabornological, and therefore quasi-ultrabarreled.

Let ν be any continuous $\mathfrak{J}$-semi-norm for this topology. Let $\rho(t) = \sup_{i \in \mathbb{N}} \nu(t\, u_i)$, then ρ is subadditive. And $\rho(t) \to 0$ as $t \to 0$ because $S = \{u_i\}$ is bounded. We can associate to ρ a continuous $\mathfrak{J}$-semi-norm ν_ρ on $\mathbb{R}^{(\mathbb{N})}$, putting $\nu_\rho(\Sigma\, t_i\, u_i) = \Sigma \rho(|t_i|)$, then $\nu \leqslant \nu_\rho$ In other words, we have found a fundamental system of -semi-norms for this topology.
Clearly, these $\mathfrak{J}$-semi-norms determine on $\mathbb{R}^{(\mathbb{N})}$ a complete vector space topology, so $\mathbb{R}^{(\mathbb{N})}$ is a complete quasi-ultrabarreled space.

Define now $T_k : \mathbb{R}^{(\mathbb{N})} \to \mathbb{R}^{(\mathbb{N})}$ by $T_k(\Sigma_0^\infty\, t_n\, u_n) = \Sigma_0^k\, n\, t_n\, u_n$.

The mappings T_k are all continuous. They are pointwise bounded, remember that only a finite number of coordinates are non zero for each $\Sigma t_n u_n \in \mathbb{R}^{(\mathbb{N})}$. Yet these mappings are not equicontinuous. So the topology we are considering is not ultrabarreled.

Notes and remarks : The contents of paragraph 1 are standard by now, and can be found in most treatises on topological vector space theory, even those that are more concerned with locally convex space theory than general topological vector space theory.

The results given in paragraph 2 are standard or trivial, except maybe proposition 3, which has been proved by S. Rolewicz [66]. Some form of proposition 3, sufficiently strong to yield its corollary 1, had earlier been proved by T. Aoki [5]. The author has not seen Aoki's original paper however, and does not know whether his results are weaker than Rolewicz's, or whether there is a misprint in the Mathematical Reviews (vol 5, 1946, p.250).

In paragraph 3, corollaries 1 and 2 of proposition 4 have been proved by S.O. Iyahen [43], and independently by J. Köhn [46].

Ultrabarreled spaces have been defined by W. Robertson [65] as topological vector spaces $(E, \mathcal{T})$ such that $\mathcal{T}' \leq \mathcal{T}$ when $\mathcal{T}'$ is a vector space topology on E in which the origin has a fundamental system of $\mathcal{T}$-closed neighbourhoods. She proved that these topological vector spaces had the Banach-Steinhaus property, and that a closed linear map of an ultrabarreled space into a Fréchet space was continuous.

S. O. Iyahen proved that $(E, \mathcal{T})$ was ultrabarreled if every closed linear map of E into a Fréchet space was continuous, and studied the ultrabornological spaces, and the quasi-ultrabarreled spaces.

Lemma 2, paragraph 4 is a rewording of his lemma 3.1 ([43], p.294).

W. Robertson, J. Köhn, S. O. Iyahen describe classes of spaces which have the "Banach-Steinhaus property".
It does not seem that any of these authors show that a topological vector space with the Banach-Steinhaus praperty was ultrabarreled; or that a space is σ-ultrabarreled if $\{u_i\}$ is equicontinuous as soon as $\cup_i u_i S$ is bounded for all $S \in \sigma$, so these results would be new.

CHAPTER II

Completant bounded structures

It is a generally recognized fact that vector space and algebra topologies are not always adapted to the study of problems involving topological algebras. Many important, useful algebras, do not have a reasonable, jointly continuous multiplication. And if $\mathfrak{a}$ is a commutative algebra and $x.y$ is not jointly continuous, then x^2 cannot be continuous because $2\,x.y = (x + y)^2 - x^2 - y^2$.

How does one prove theorems then ? It is often possible to find another structure on the algebra, with respect to which the algebraic operations behave well. If the relations between the original topology and the new structure are narrow enough, it will be possible to translate the topological data, handle these in the new structure, and translate the results back into topological language.

Completant bounded structures constitute a class of structures with respect to which the author has often been able to carry out this program successfully.

1. <u>Boundednesses</u> : <u>Definition 1</u> : An absolutely convex subset of a real or a complex vector space E is norming if it does not contain any non zero vector subspace.

Its Minkowski functional is then a norm.

<u>Definition 2</u> : A subset of a real or of a complex vector space is completant if it is norming, and its Minkowski functional is a Banach space norm.

<u>Definition 3</u> : A bounded structure on a set E is a set $\mathcal{B}$ of subsets of E such that

a) $B_1 \cup B_2 \in \mathcal{B}$ if $B_1 \in \mathcal{B}$, $B_2 \in \mathcal{B}$;

b) $B' \in \mathcal{B}$ if $B' \subseteq B \in \mathcal{B}$,

c) E is the union of all the elements of $\mathcal{B}$. These elements of $\mathcal{B}$ are called the bounded sets of the boundedness or bounded structure.

In other words, the set of complements of bounded sets is a filter with empty intersection.

Definition 4 : Let E be a real or a complex vector space. A bounded structure on E is a vector space bounded structure if
a) every bounded set B is contained in some bounded balanced set B',
b) a finite sum of bounded sets is bounded.

Definition 4' : Assume E is furthermore fitted out with a multiplication which makes E into an algebra. The boundedness $\mathcal{B}$ on E is an algebra boundedness if it is a vector space boundedness and
c) the product of two bounded sets is bounded.

Definition 5 : A vector space bounded structure on a real or a complex vector space is a convex bounded structure if every bounded set is contained in a convex bounded set. It is a norming bounded structure if every bounded set is contained in a norming bounded set. It is a completant bounded structure if every bounded set is contained in a completant bounded set.

We shall often be considering vector spaces fitted out with completant bounded structures, and shall call these b-spaces.

Definition 6 : Let $(E,\mathcal{B})$, $(F,\mathcal{C})$ be two sets on which bounded structures are defined. A mapping $u : E \to F$ is a morphism if $uB \in \mathcal{C}$ whenever $B \in \mathcal{B}$. If E, F are vector spaces, and $\mathcal{B}, \mathcal{C}$ are vector space bounded structures, $u : E \to F$ is a morphism if u is linear, and a morphism for the bounded structures.
The space of morphisms of one vector space with boundedness into another is clearly a vector subspace of the space of all linear mappings.

Definition 7 : Let $(E,\mathcal{B})$, $(F,\mathcal{C})$ be two sets with boundedness. A set B_1 of mappings $u : E \to F$ is equibounded if

$$\bigcup_{u \in B_1} uB$$

is a bounded subset of F whenever B is a bounded subset of E.

The set of equibounded sets of mappings is a boundedness on the set of all morphisms $(E,\mathcal{B}) \to (F,\mathcal{C})$. If E, F are vector spaces,

and $\mathcal{B}, \mathcal{C}$ are vector space boundednesses, the equibounded bound ness is a vector space boundedness on the space of morphisms of ve spaces with boundednesses of E into F. The equibounded boundedn is convex, norming, or completant respectively when the boundednes F is convex, norming, or is completant.

Definition 8 : Let $\mathcal{B}_1$ and $\mathcal{B}_2$ be two bounded structures on E We say that $\mathcal{B}_1$ is finer (stronger) than $\mathcal{B}_2$ if $\mathcal{B}_1 \subseteq \mathcal{B}_2$, i. if the identity mapping $(E, \mathcal{B}_1) \to (E, \mathcal{B}_2)$ is a morphism.

2. Boundednesses and topologies : We want to use the bounded struc as tools in the study of topological vector spaces and topologica gebras. If $(E, \mathcal{T})$ is a topological vector space, a subset $B \subseteq$ is bounded (topologically) if we can associate to every neighbour U of the origin an $\varepsilon > 0$ such that $\varepsilon B \subseteq U$ when $|\lambda| \leqslant \varepsilon$.

It is well known that this is a vector space bounded structure.

Definition 9 : The set of all topologically bounded subsets of E the topological bounded structure of E, if $(E, \mathcal{T})$ is a topolog vector space.

Definition 10 : A vector bounded structure $\mathcal{B}$ and a vector topol $\mathcal{T}$ are compatible on a vector space E if $\mathcal{B}$ is stronger than topological bounded structure of E.

Definition 11 : The convex boundedness of a topological vector sp E is the set of the bounded sets $B \subseteq E$ which have a bounded c hull.

This is the weakest boundedness which is convex, and stro than the topological boundedness. It is norming if the topologica vector space E is Hausdorff. The convex boundedness of a vector space E with a general, vector space boundedness, could be defi just as easily. We shall probably not need it.
The following result is often useful.

Proposition 1 : The convex boundedness of a Hausdorff topological tor space E is completant if and only if the closed absolutely vex, bounded subsets are completant. And the condition will be sa fied if and only if a Cauchy sequence of a normed space E_B, B

vex and bounded in E has a limit in E.

The convex boundedness is trivially completant when closed absolutely convex bounded sets are. When this second condition is satisfied, a Cauchy sequence of E_B has a limit in E_B, and therefore in E. (As usual E_B is the vector space generated by the bounded set B, normed by the Minkowski functional of B). To end the proof we must show that :

<u>Proposition 2</u> : A bounded closed absolutely convex subset B of a topological vector space E is completant if the Cauchy sequences of E_B have limits for the topology of E.

Let x_n be a Cauchy sequence of E_B, we may assume that $x_n \in B$ without essential loss of generality, and assume that $x_n \to x$ for the topology of E. Then $x \in B$ because B is closed. Also $x_n - x_m \in \varepsilon_n B$ when $m \geq n$ with $\varepsilon_n \to 0$, so at the limit $x_n - x \in \varepsilon_n B$.

Assume that a sequence ε_n of strictly positive numbers can be found in such a way that for every neighbourhood U of the origin a neighbourhood V can be found, such that

$$U \supseteq \bigcup_N \Sigma_1^N \varepsilon_k V$$

<u>Proposition 3</u> : When this is the case, the convex boundedness of E is completant if and only if $\Sigma \lambda_k b_k$ converges for every bounded sequence b_k and every sequence $\lambda_k = o(\varepsilon_k)$.

The condition is sufficient. Let B be a bounded, closed, absolutely convex subset of E. A Cauchy sequence of E_B has a subsequence x_k such that $x_{k+1} - x_k \in \varepsilon_k B$. The condition implies that x_k converges.

It is necessary. Let b_k be a bounded sequence, and $\lambda_k = o(\varepsilon_k)$. Consider the set

$$B = \left\{ \Sigma_1^N \mu_k b_k \;\middle|\; \forall k : |\mu_k| \leq \lambda_k, N \in \mathbb{N} \right\}$$

It is a bounded absolutely convex set. The partial sums of the series $\Sigma \mu_k x_k$ are a Cauchy sequence of the normed space E_B.

Proposition 4 : Let A be an algebra, with a locally convex topology defined in such a way that multiplication is separately continuous, the topological boundedness of A being furthermore assumed completant. Multiplication is then bounded in A.

This follows from the uniform boundedness principle. We must show that $B_1.B_2$ is bounded when B_1 and B_2 are. We may assume without loss of generality that B_1 and B_2 are completant. Multiplication is then separately continuous $A_{B_1} \times A_{B_2} \to A$, and hence continuous. And $B_1.B_2$ is bounded.

This result is not easy to generalize. If the algebra is not locally convex, or if it is locally convex, but if its convex boundedness is not completant, we would still be able to show that the product of two bounded completant sets is bounded (even more, but never mind). The product would however have no reason to be contained in any bounded completant set, so that we would not get a multiplicative boundedness in this way.

Something can be done in the locally pseudo-convex case, however, when we introduce the "small boundedness".

Definition 12 : A sequence of elements x_k of a topological vector space is rapidly decreasing if $k^n x_k$ is a bounded sequence for all values of n. A subset of a locally pseudo-convex space is a small bounded subset if it is contained in the closed absolutely convex hull of a rapidly decreasing sequence.

It is easy to see that the small bounded subsets of a locally pseudo-convex space are bounded. Let x_k be a rapidly decreasing sequence, and ν a p-semi-norm.

We must see that ν is bounded on the convex hull of $\{x_k\}$, but

$$\nu(\Sigma\, \lambda_k\, x_k) \leq \Sigma\, |\lambda_k|^p\ k^{-np}\ \sup_k\, \nu\,(n^k\, x_k)$$

We choose n in such a way that $np > 1$, $\Sigma\, k^{-np}$ is then finite, the λ_k are all smaller than unity in absolute value, and $\nu(n^k\, x_k)$ is bounded since x_k is a rapidly decreasing sequence.

Proposition 5 : Let A be an algebra, on which a locally pseudo-convex topology is defined. Assume that the convex boundedness of A is completant, and that multiplication is separately continuous. The product of two small bounded sets is a small bounded set.

So, the small boundedness of A is multiplicative under these hypotheses. What has to be proved is that $\{x_k \; y_{k'}\}$ can be ordered in such a way that it becomes a rapidly decreasing sequence, if we assume that the given sequences $\{x_k\}$, $\{y_{k'}\}$ are rapidly decreasing.

We order the set of couples (k, k') by increasing values of $k.k'$, defining thus some function $\ell(k,k')$, bijective between $\mathbb{N} \times \mathbb{N}$ and $\mathbb{N}$. Then $\ell(k,k') \leq k.k' \,(1 + \log k.k')$ - just count the number of couples (k,k') with $k.k' \leq \ell$. Put $z_{\ell(k,k')} = x_k y_{k'}$.

The sequences $k^n \, x_k$ and $k'^n \, y_{k'}$ are contained in bounded completant sets. The product of these sequences is therefore bounded, i.e. $(k.k')^n \; x_k \, y_{k'}$ is a bounded sequence, hence $\ell^{n-\varepsilon} \; z_\ell$ is also bounded. This ends the proof.

It is not often that we shall be dealing with a locally convex algebra, whose topological boundedness will be easy to determine. We shall however often be considering algebras of operators. The equicontinuous boundedness is a very natural object on such an algebra.

Proposition 6 : Let E be a locally convex space. The set of equicontinuous sets of linear transformations is an algebra boundedness on $\mathcal{L}(E, E)$, the algebra of continuous linear mappings of E into itself. If the closed absolutely convex bounded subsets of E are completant, then $\mathcal{L}(E, E)$ is a b-algebra with this boundedness.

This is trivial. When E is a locally pseudo-convex space, we meet the following hitch, that the equicontinuous boundedness of $\mathcal{L}(E, E)$ is in general not even pseudo-convex.
Consideration of "small bounded sets" may however save us :

Definition 13 : A sequence $\{u_k\}$ of linear transformations of a locally pseudo-convex space E is rapidly decreasing if $k^n \, u_k$ is an equicontinuous set of transformations for all n. A set of linear transformations of E is a small equicontinuous set if it is contained in the pointwise closure of the absolutely convex hull of a rapidly decreasing sequence.

We must see that small equicontinuous sets are equicontinuous. Let u_k be a rapidly decreasing sequence of linear transformations of E, let U be a neighbourhood of the origin in F, we may assume that $U = \{x \mid \nu(x) \leq 1\}$ for some p-semi-norm ν. Choose n with $np > 1$, then V such that $k^n u_k V \subseteq U$, V a neighbourhood of the origin in E. Let f be in the absolutely convex hull of the u_k, then $f V \subseteq M U$ where $M = (\Sigma k^{-np})^{1/p}$ is a finite number.

This shows that the absolutely convex hull of a rapidly decreasing sequence is equicontinuous. The pointwise closure of an equicontinuous set of linear transformations is also equicontinuous, of course.

Proposition 7 : The small equicontinuous boundedness of the algebra of linear transformations of a locally pseudo-convex space is multiplicative. With this boundedness, $\mathcal{L}(E, E)$ is a b-algebra when the closed absolutely convex subsets of E are completant.

This should be clear by now. We must show that $u_k \circ v_{k'}$ can be reordered in such a way that it becomes rapidly decreasing if u_k and $v_{k'}$ are rapidly decreasing sequences. The reordering is the same as the one considered in the proof of proposition 5.

3. Subspaces. Closed subspaces : Let $(E, \mathcal{B})$ be a b-space. The following object will be interesting. F will be a vector subspace of E, $\mathcal{C}$ a b-space boundedness on F, the elements of $\mathcal{C}$ belonging all to $\mathcal{B}$. We require thus that the bounded subsets of F be bounded in E, but do not require the bounded subsets of E which are contained in F to be bounded in F.

Definition 14 : When these conditions hold, we say that $(F, \mathcal{C})$ is a b-subspace of $(E, \mathcal{B})$.

Let $(E_1, \mathcal{B}_1)$ and $(E_2, \mathcal{B}_2)$ be two b-spaces, and $u : E_1 \to E_2$ a morphism. Let $u\mathcal{B}_1 = \{uB \mid B \in \mathcal{B}_1\}$. Then $(u E_1, u\mathcal{B}_1)$ is a b-subspace of $(E_2, \mathcal{B}_2)$.

We need only check that uB is completant when B is completant. Of course uB is absolutely convex. It is a bounded subset of F and must therefore be norming. The result will follow from

Proposition 8 : The linear image uB of a completant set B is completant when it is norming. It is norming if and only if $B \cap \operatorname{Ker} u$ is completant.

This follows from the fact that F_{uB} can be identified, as a normed or a seminormed space with the quotient $E_B/E_{B \cap \operatorname{Ker} u}$. The quotient is a Banach space if it is a normed space. It is a normed space when $E_{B \cap \operatorname{Ker} u}$ is a Banach space.

Definition 15 : The b-space $(u E_1, u\mathcal{B}_1)$ is the image of $(E_1, \mathcal{B}_1)$ by the morphism u.

We shall also be interested in kernels of morphisms. We shall put the induced boundedness on such a kernel - a subset of a kernel will be called bounded if it is bounded in the original space. With such a boundedness, a kernel will be called a closed subspace.

Definition 16 : Let $(E, \mathcal{B})$ be a b-space, and $(F, \mathcal{C})$ a subspace of $(E, \mathcal{B})$, which is a b-space, and is such that every bounded subset of E which is contained in F is a bounded subset of F. Then $(F, \mathcal{C})$ is a closed subspace of $(E, \mathcal{B})$.

We must see that this definition sticks with the general description of a closed subspace given above, i.e. we must show that the kernel of a morphism is a closed subspace, define the quotient of a b-space with respect to a closed subspace and show that the quotient is a b-space. The induced boundedness is trivially convex, and norming. To show it is completant, we must see that $B \cap \operatorname{Ker} u$ is completant when B is completant and u is b-space morphism. This follows from proposition 11 since uB is norming.

Conversely, assume that F is a closed subspace. Consider the space E/F, put on E/F the quotient boundedness : a subset of E/F would be bounded if it is the image of a bounded subset of E by the quotient morphism.

It is trivial that the quotient boundedness is convex. It is completant if it is norming, to show it is norming, it is sufficient to show that $B \cap F$ is completant when B is a completant subset of E, again by proposition 11.

$B \cap F$ is a bounded subset of F, so a bounded completant subset B_1 of F can be found, which contains $B \cap F$. Then $B \cap F = B \cap B_1$. To show that $B \cap B_1$ is completant, we apply again proposition 11, to the mapping $(x,y) \to x - y$ of $F \times F_1$ into F. The kernel of this mapping is the diagonal. Since $B - B_1$ is norming, we see that $B \times B_1$ intersects the diagonal along a completant set, i.e. $B \cap B_1$ is completant.

Note : The reader must be explicitly aware of the fact that we are "cheating". The b-spaces are, in a way, complete. We defined a "closed" subspace as one whose induced boundedness is a b-space boundedness. The fact that a subspace is the kernel of a morphism if, and only if its induced boundedness is completant may compared to the theorem that a subgroup of a complete group is a kernel if, and only if its induced topology is complete.

4. On the closure of a subspace : It is clear that any intersection of closed subspaces of a b-space is closed. The closure of a (not necessarily closed) subspace of a b-space is the intersection of the closed subspaces containing it. It is to be expected that the closure of a subspace can be described by a system of limiting operations. The limiting operations being sequential, and a b-space not being a topological vector space, it is not surprising that this construction is not as simple as in the topological case.

Definition 17 : Let E be a b-space, x_n a sequence of elements of E, and x an element of E. Then $x_n \to x$ if a sequence ε_n of positive numbers tending to zero can be found in such a way that $(x_n - x)/\varepsilon_n$ is a bounded sequence. The sequence x_n is a Cauchy sequence if the $\varepsilon_n \to 0$ can be found in such a way that $\{(x_m - x_n)/\varepsilon_n \mid m \geq n\}$ is a bounded set in E.

It is clear that Cauchy sequences converge : choose B bounded and completant, and such that $x_m - x_n \in \varepsilon_n B$ when $m \geq n$. The sequence x_n is then a Cauchy sequence of the Banach space E_B.

Let now B be a bounded, absolutely convex subset of the space E. Let B_1 be the set of all $\lim x_n$ where all $x_n \in B$, and $x_n - x_{n+1} \in 2^{-n}B$ - this condition implies that the sequence x_n is Cauchy in the normed space E_B, every Cauchy sequence in E_B has a subsequence such as x_n. Then B_1 is completant, it is the unit ball of a quotient of the completion of E_B.

Instead of B_1, we could have considered B_1', the intersection of all bounded completant B' which contain B. This is again a completant set, an intersection of bounded completant sets is completant : $\bigcap_{i \in I} B_i$ is the unit ball of the kernel of the linear mapping $X_C \to Y_D$ where $C = \Pi_i B_i$, $D = \Pi_{i,j} (B_i + B_j)$, where X_C and Y_D are the linear spaces generated by C and D, and where our linear mapping maps (x_i) onto $(x_i - x_j)$. There is not much difference between B_1 and B_1', a little thought will show that they are two absolutely convex sets with the same Minkowski functional. Either can be called the completant hull of B.

<u>Proposition 9</u> : A subspace F of a b-space E is closed iff all convergent sequences of F have limits in F.

We must check that the induced boundedness of F is completant when all convergent sequences of F have limits in F, and conversely, that all convergent sequences of F have limits in F when the induced boundedness of F is completant.

Assume first that the limits of convergent sequences are in F. Let B_0 be a bounded subset of F, then B its absolutely convex hull. Then $B \subseteq F$ and B is a bounded subset of E. Let B_1 be the completant hull of B, then B_1 is bounded in E, is completant, the elements of B_1 are limits of elements of F, hence $B_1 \subseteq F$. Hence the induced boundedness of F is completant.

Conversely, assume the induced boundedness of F is completant, and consider a sequence of elements of F which has some limit in E. The sequence is Cauchy in F for the induced boundedness. It must therefore have some limit in F, i.e. the limit of the sequence is in F.

We can now describe the system of limiting operations which lead from a subspace to its closure.

<u>Definition 18_1</u> : Let E be a b-space, F a vector subspace of E. We shall let B_1 be the completant hull of B for each bounded absolutely convex subset B of E. We also let

$$\lim F = \bigcup_B B_1$$

where B ranges over the bounded absolutely convex subsets of E which are contained in F.

$\lim F$ carries a natural completant boundedness. The set $C \subseteq \lim F$ is bounded for its boundedness if $C \subseteq B_1$ for some absolutely convex bounded subset B of E with $B \subseteq F$. With this boundedness, $\lim F$ is a b-space.

It is not necessarily a closed b-subspace of E. There may be bounded subsets of E which are contained in $\lim F$ but are not contained in B_1 for any bounded subset B of E. We therefore forget the natural boundedness of $\lim F$ and try again, transfinitely.

<u>Definition 18_α</u> : Assume that α is an ordinal. If α is not a limit ordinal, $\alpha = \beta + 1$. We define $\lim_\alpha F = \lim \lim_\beta F$ if $\lim_\beta F$ has been defined. If α is a limit ordinal, we define $\lim_\alpha F = \bigcup_{\beta<\alpha} \lim_\beta F$ if $\lim_\beta F$ has been defined for all $\beta < \alpha$.

All these spaces are, in a natural way, b-spaces. The boundedness of $\lim_\alpha F = \lim F_\beta$ with $F_\beta = \lim_\beta F$ is defined as in definition 18_1, replacing F by F_β. The boundedness of $\bigcup_{\beta<\alpha} F_\beta$ is the union of the bounded structures of the several F_β.

It is clear, for set theoretical reasons, that some ordinal α will be reached such that $\lim_\alpha F = \lim_{\alpha+1} F$. And the vector space obtained in this way must be closed, and be contained in every closed vector space containing F, i.e. this $\lim_\alpha F$ is the closure of F.

<u>Proposition 9</u> : The first ordinal for which $\lim_\alpha F$ is the closure of F is at most the first uncountable ordinal, say ω_1.

Let B be a completant subset of E. We shall show that $B \cap \lim_{\omega_1} F$ is completant. Consider a Cauchy sequence, $x_1, \ldots, x_n, \ldots$ of $E_{B \cap \lim_{\omega_1} F}$, this is a Cauchy sequence of E_B whose elements also belong to $\lim_{\omega_1} F$. Each individual x_n belongs to some $\lim_{\alpha_n} F$, the ordinals α_n are countable and have a countable upper bound.

We are therefore considering a Cauchy sequence of E_B whose elements belong to $\lim_\alpha F$. The limit must be in $\lim_{\alpha+1} F \subseteq \lim_{\omega_1} F$. Our Cauchy sequence therefore converges.

<u>Proposition 10</u> : Consider the space $C(I)$ of all continuous functions on the unit interval, in the space $\mathbb{C}^I$ of all mappings $I \to \mathbb{C}$, with the product boundedness - a set $B \subseteq \mathbb{C}^I$ is bounded if all of the coordinate projections (evaluation functions) $\mathbb{C}^I \to \mathbb{C}$ map B onto bounded subsets of $\mathbb{C}$. The space $\lim_\alpha C(I)$ is then not closed for any countable α.

We shall define Baire classes of functions, putting $C_0(I) = C(I)$ letting $C_{\alpha+1}(I)$ be the set of all pointwise limits of elements of $C_\alpha(I)$, and when α is a limit ordinal, letting $C_\alpha(I) = \bigcup_{\beta<\alpha} C_\beta(I)$. Baire ([77], chap. VIII, ¶ 5) has shown that $C_{\omega_1}(I)$ was the space of all Borel measurable functions, but that $C_\alpha(I) \neq C_{\omega_1}(I)$ for all $\alpha < \omega_1$.

It is clear that $\lim_\alpha C(I) \subseteq C_\alpha(I)$. So proposition 3 will follow once we show that the closure of $C(I)$ is the space of all Borel functions. It will clearly be sufficient to prove that the characteristic function of every Borel set is in this closure, i.e. that the set of $K \subseteq I$ such that $1_K \in \overline{C(I)}$ is a σ-ring which contains the half-open intervals $]a,b]$, $a < b$.

We let f be some continuous function on the interval $[0,1]$, $f = 1$ on $[0,a]$ and $0 \leq f < 1$ on $]a,1]$.
We also let $\varphi = 1$ on $[0,a]$, $\varphi = 1 - f$ on $]a,1]$.
Then φ never vanishes. The set B_φ of functions u such that $\varphi|u| \leq 1$ is an absolutely convex bounded set of $C(I)$ whose Minkowski functional is $\sup_t \varphi(t)\,|u(t)|$.
We observe that $\varphi.f^n \to \varphi.1_{[0,a[}$ when $n \to \infty$, uniformly, and deduce from that fact that $f^n \to 1_{[0,a[}$ for the Minkowski functional of B_φ. This proves already that $1_{[0,a[} \in \overline{C(I)}$, hence

$1_{]a,b[} = 1_{[o,b[} - 1_{[o,a[}$ also belongs to $C(I)$.

Because $\mathbb{C}^I$ is a b-algebra, and $C(I)$ a subalgebra, it follows that $\overline{C(I)}$ is a b-algebra, so if U, V are sets such that $1_U \in \overline{C(I)}$, $1_V \in \overline{C(I)}$, then $1_{U \cap V} = 1_U \cdot 1_V \in \overline{C(I)}$, and putting $U \Delta V = (U \cup V) \setminus (U \cap V)$, then $1_{U \Delta V} = 1_U + 1_V - 2\, 1_U\, 1_V \in \overline{C(I)}$, i.e. the class of sets we are considering is an algebra of sets.

We must still verify that if U_n is an increasing sequence of sets, and $\forall n : 1_{U_n} \in \overline{C(I)}$, then $1_U \in \overline{C(I)}$ where $U = \bigcup U_n$. But, let now $\varphi(t) = 1$ on U_1, $\varphi(t) = 1/n$ on $U_{n+1} \setminus U_n$, and $\varphi(t) = 1$ on the complement of U. Again $\varphi(t) \cdot 1_{U_n} \to \varphi(t) \cdot 1_U$ uniformly, and that is the desired result.

5. <u>b-spaces of type $\mathfrak{F}$ and $\mathcal{L}\mathfrak{F}$</u> : An $\mathfrak{F}$-space, or a Fréchet space, is a complete, metrizable topological vector space. We do not assume $\mathfrak{F}$-spaces to be locally convex.

<u>Definition 19</u> : A b-space E is of type $\mathfrak{F}$ if its boundedness is the convex boundedness associated to a topology of type $\mathfrak{F}$ on E.

The convex boundedness of a space of type $\mathfrak{F}$ is obviously completant. Let us show that the boundedness of a b-space of type $\mathfrak{F}$ is sufficient to recover its topology.

<u>Proposition 11</u> : Let $(E, \mathcal{T})$ be a complete metrizable topological vector space, and let $\mathcal{B}$ be its convex boundedness. Then $\mathcal{T}$ is the strongest topology on E which induces a weaker topology than the norm-topology on each E_B, B bounded completant.

<u>Remark 1</u> : The proof of proposition **11** goes through when $(E, \mathcal{T})$ is any metrizable topological vector space, and $\mathcal{B}$ is its convex boundedness. $\mathcal{B}$ is not a completant boundedness when $\mathcal{T}$ is not complete, because a Cauchy sequence for $(E, \mathcal{T})$ has a subsequence which is Cauchy for $(E, \mathcal{B})$. There are examples where $\mathcal{T}$ is not the direct limit of the spaces E_B, with B bounded and completant, yet $\mathcal{T}$ is a metrizable vector space topology on E, see further, remark 3.

<u>Remark 2</u> : We may replace the convex boundedness considered in proposition 11 by a suitable, stronger boundedness, e.g. the set of B'

with B' precompact in some E_B, B absolutely convex bounded (this is the weakest boundedness of type ($\mathfrak{S}$) stronger than $\mathfrak{B}$) or again by the subsets of the closed absolutely convex hulls of sequences (x_k) where $k^n x_k$ has a bounded absolutely convex hull for all n (this is the weakest nuclear boundedness stronger than $\mathfrak{B}$), or again by many bounded structures whose definitions involve rapid convergence of structures.

The proof of such results uses the following triviality :
Let E be a normed space. Let $\{\lambda_n\}$ be a sequence of strictly positive numbers tending to zero.
Let $\mathfrak{B}$ be the boundedness generated by the convex hulls of sequences (x_n) where x_n/λ_n is bounded. The norm topology of E is the strongest which induces for all $B \in \mathfrak{B}$, absolutely convex, a weaker topology than that of the Minkowski functional of B.

To prove proposition 11, we observe that the strongest topology $\mathfrak{T}$, on E, which induces a weaker topology than the norm topology on E_B, for all B bounded completant is clearly translation and homothetically invariant. If it is strictly stronger than $\mathfrak{T}$, we can find a sequence x_k which tends to zero in $(E,\mathfrak{T})$, and a neighbourhood V of the origin in $\mathfrak{T}_1$ such that no x_k belongs to V.

The sequence x_k has a subsequence y_k such that $y_k \to 0$ bornologically. Just choose a fundamental sequence $\{U_k\}$ of neighbourhoods of the origin for $\mathfrak{T}$ such that $U_k \supseteq U_{k+1} + U_{k+1}$, and choose $y_k \in 2^{-k} U_k$. The closed absolutely convex hull B of the sequence $2^k y_k$ is a bounded completant subset, and $y_k \to 0$ in E_B.
So $y_k \to 0$ in $(E,\mathfrak{T}_1)$, which contradicts the hypothesis that x_k remains out of V.

We have used the fact that the closed absolutely convex hull of (y_k) is bounded if $y_k \in U_k$, where $\{U_k\}$ is a fundamental sequence of balanced neighbourhoods of the origin in $(E,\mathfrak{T})$, and $U_k \supseteq U_{k+1} + U_{k+1}$. This closed absolutely convex hull is even compact. Consider the series $\Sigma\, \lambda_k\, y_k$, where λ_k ranges over the unit ball of ℓ_∞. It converges uniformly because

$$\Sigma_{k+1}^{k'} \lambda_n\, y_n \in U_k$$

so the mapping $(\lambda_k) \to \Sigma\, \lambda_k\, y_k$ is a $\sigma(\ell_\infty, \ell_1)$-continuous mapping of the unit ball of ℓ_∞ into our topological vector space. The image is compact, convex, and contains the given sequence.

Definition 20 : Let $E_1, \ldots, E_n, \ldots$ be an increasing sequence of vector spaces. For each n, let $\mathcal{B}_n$ be a bounded structure of type $\mathfrak{F}$ on E_n. Assume that $\mathcal{B}_n \subseteq \mathcal{B}_{n+1}$. Let $E = \bigcup E_n$, $\mathcal{B} = \bigcup \mathcal{B}_n$. Then $(E,\mathcal{B})$ is a b-space of type $\mathcal{L}\mathfrak{F}$.

Proposition 12 : Let $(E,\mathcal{B})$ be a b-space of type $\mathcal{L}\mathfrak{F}$, so that $E = \bigcup E_n$, $\mathcal{B} = \bigcup \mathcal{B}_n$ with $(E_n,\mathcal{B}_n)$ a b-space of type $\mathfrak{F}$. Let $(F,\mathcal{C})$ be a b-space of type $\mathfrak{F}$, and $u : F \to E$ a linear mapping with closed graph. Then $u F \subseteq E_n$ for n large, and for that value of n, $u \mathcal{C} \subseteq \mathcal{B}_n$.

The fact that $u \mathcal{C} \subseteq \mathcal{B}_n$ as soon as $u F \subseteq E_n$ follows from Banach's closed graph theorem. The spaces F and E_n are Fréchet space, $u : F \to E_n$ has a graph which is closed for the bounded structure of $F \times E_n$. Proposition 11 applied to the Fréchet space $F \times E_n$ shows that the graph is then closed for the topology of $F \times E_n$. So u is continuous, and a fortiori a b-space morphism.

Ker u is the set of $x \in F$ with $(x,0)$ in the graph of u. It is the intersection of two closed spaces and therefore closed (in $F \times E$, but we embed $F \subseteq F \times E$ mapping x on $(x,0)$). So $F/\text{Ker } u$ is a new b-space of type $\mathfrak{F}$, say $F/\text{Ker } u = F_1$, u induces a mapping $u_1 : F_1 \to E$. We identify F_1 with $u F_1$, and see that it is sufficient to prove proposition 12 when $F \subseteq E$ and u is the identity mapping. The diagonal, the set of (x,x), $x \in F$, is a closed subspace of $F \times E$ since u has closed graph.

We shall let $F_n = E_n \cap F$, and put on F_n the upper bound of the topologies of E_n and F. This is a Fréchet topology, it is the topology of the diagonal in $E_n \times F$, and the diagonal is closed for the boundedness of $E_n \times F$, hence for the Fréchet topology of $E_n \times F$ too.

The Fréchet topology of F will be called $\mathcal{T}$. If $X \subseteq F$, $\overline{X}$ will be the closure of X for $\mathcal{T}$ (rather than any $\mathcal{T}_n$).

Let $U_{n.1}$ $U_{n.2}$ ... be a fundamental system of balanced neighbourhoods of the origin in F_n, and assume that $U_{n.k} \supseteq U_{n.k+1} + U_{n.k+1}$. Let $k_1, \ldots, k_n, \ldots$ be an arbitrary sequence of integers. E is the union of the sets $M\, U_{n.k_n+1}$, $M \in \mathbb{R}_+$, $n \in \mathbb{N}$.
Baire's theorem shows that at least one of the sets $\overline{U_{n.k_n+1}}$ has an interior, and $\overline{U_{n.k_n}}$ must be a neighbourhood of the origin (for the topology $\mathcal{T}$).

It is possible to find n such that all $\overline{U_{n.k}}$ are neighbourhoods of the origin, since otherwise we would have for each n some k_n such that $\overline{U_{n.k_n}}$ is not a neighbourhood, and no such sequence of integers exists.

The end of the proof is straightforward, we must show that $F = F_n$, if F_n has a stronger topology than F, both being Fréchet spaces, and the closure of each neighbourhood of the origin in F_n is a neighbourhood of the origin in F. We choose $x \in F$, then $x \in \overline{F_n}$. We write $x = y_1 + x_1$ with $y_1 \in F_n$, $x_1 \in 2^{-1}\ \overline{U_{n.1}}$.
More generally, if $x_k \in 2^{-k}\ \overline{U_{n.k}}$ we write $x_k = x_{k+1} + y_{k+1}$ where $x_{k+1} \in 2^{-k-1}\ \overline{U_{n.k+1}}$, $y_{k+1} \in 2^{-k}\ U_{n.k}$. We see that $\Sigma\ y_k$ is a convergent series in F_n, but $x_k \to 0$ in F so $x = \Sigma\, y_k \in F_n$.
Hence $F = F_n \subseteq E_n$.

Corollary 1 : Let $(E,\mathcal{B})$ be a b-space of type $\mathcal{LF}$, $E = \bigcup E_n$, $\mathcal{B} = \bigcup \mathcal{B}_n$, each $(E_n, \mathcal{B}_n)$ being a b-space of type $\mathcal{F}$. Let $(F,\mathcal{C})$ be a b-space of type $\mathcal{F}$, where $F \subseteq E$, $\mathcal{C} \subseteq \mathcal{B}$. Then $F \subseteq E_n$ for some large n, and $\mathcal{C} \subseteq \mathcal{B}_n$ for that value of n.

This is trivial by now ; since $\mathcal{C} \subseteq \mathcal{B}$, the identity $(F,\mathcal{C}) \to (E,\mathcal{B})$ is a morphism and therefore has closed graph. So the corollary is hardly worth mentioning. It shows however that $(E,\mathcal{B})$ is a direct limit of Fréchet spaces in an essentially unique way : If $E = \bigcup E'_n$, $\mathcal{B} = \bigcup \mathcal{B}'_n$, $(E'_n, \mathcal{B}'_n)$ of type $\mathcal{F}$ is another description of $(E,\mathcal{B})$ as an $\mathcal{LF}$ space, then a sequence $k_1, \ldots$ of integers can be found such that $E_n \subseteq E'_{k_n}$, $\mathcal{B}_n \subseteq \mathcal{B}'_{k_n}$, $E'_n \subseteq E_{k_n}$, $\mathcal{B}'_n \subseteq \mathcal{B}_{k_n}$.

Corollary 2 : A b-structure of type $\mathcal{L}\mathfrak{F}$ is maximal among completant bounded structures.

Let $(E,\mathfrak{B})$ be a b-space of type $\mathcal{L}\mathfrak{F}$, let $\mathfrak{B}' \supseteq \mathfrak{B}$ be a larger completant boundedness, and $B \in \mathfrak{B}'$ be completant. The identity mapping $E_B \to E$ has then a closed graph : the graph boundedness on the diagonal (identified with E_B) is generated by the sets $M B \cap B_1$ with $M \in \mathbb{R}_+$, $B_1 \in \mathfrak{B}'$, B_1 completant, and $M B \cap B_1$ is completant since $M B + B_1$ is norming. So the induced boundedness of the graph is completant.

Corollary 3 : (Buchwalter). A b-structure of countable type is maximal.

A b-space is of countable type if there exists a fundamental sequence of bounded sets. Such a b-space is a countable direct limit of Banach spaces, and is therefore an $\mathcal{L}\mathfrak{F}$-space.

Remark 3 : We promised to give examples where a metric vector space topology $\mathfrak{T}$ on a vector space was not the direct limit of the topologies E_B, B bounded and completant.

Let $0 < p < 1$, consider the space ℓ_p, with the norm $\| \ \|_1$. Let $\mathfrak{B}_1$ be the set of $B \subseteq \ell_p$ such that some completant $B_1 \subseteq \ell_p$ can be found, with B_1 bounded in ℓ_1. Then $\mathfrak{B}_1$ is a completant boundedness on ℓ_p, and is larger than the boundedness of type $\mathfrak{F}$ which is associated to $\| \ \|_p$ on ℓ_p. So $\mathfrak{B}_1$ is that completant boundedness of type $\mathfrak{F}$.

This direct limit of the topologies E_B, B bounded and completant, is the norm topology of ℓ_p.

Let X be a compact subset of $\mathbb{C}^n$, let ν be a norm on $E = \mathfrak{O}(X)$ which induces a weaker topology on $\mathfrak{O}(X)$ than the usual direct limit topology. Again, the set of $B \subseteq \mathfrak{O}(X)$ such that $B \subseteq B_1$ with B_1 completant, and bounded for ν, is a completant boundedness which is weaker than the direct limit boundedness of $\mathfrak{O}(X)$, and this direct limit boundedness is maximal.

The direct limit of the topologies E_B, B bounded and completant in (E,ν), is the usual direct limit topology of $\mathfrak{O}(X)$.

We have the following amusing result : Let $X \subseteq \mathbb{C}^n$ be a set such that $\sup_{x \in X} |u(x)|$ is a norm on $\mathcal{O}(X)$, e.g. let X be the closure of its interior. Let $B \subseteq \mathcal{O}(X)$ be completant, and be bounded for the sup-norm. Then B is bounded in $\mathcal{O}(X)$, i.e. there is a neighbourhood U of X to which the elements of B extend the extensions being uniformly bounded.

Notes and remarks : Analysts have often considered non topological convergence structures on vector spaces. Let us not speak of the pioneering pre-Hausdorff days, when it was not clear what could (or should) be called a topology.

In his initial definition, L. Schwartz defined a convergence structure on the space $\mathcal{D}(\mathbb{R}^n)$. He replaced this later by a topology, but retained the convergence structure for pedagogical reasons in his general exposition of the theory [67].

J. Mikusinski [56] defined a field of "operators" which contains some rings of distributions on the real line (and elsewhere). He did not topologize this field but he did define a convergence structure on it [57]. His field of operators is a "union of Banach spaces", i.e. a b-space. A distribution taking its values in a union of Banach spaces is a bounded linear mapping of $\mathcal{D}(\mathbb{R}^n)$ into a b-space.

C. Foias and G. Marinescu [25] gave an abstract description of such convergence structures. They considered a more general class of spaces, and called them "polymetrizable". The b-spaces are a subclass of these, which they call "polynormed". The same authors also gave an example of a b-space which is not separated by its dual [26]. A little fiddling around this example shows that no Hausdorff, admissible locally convex topology exists on the space of Mikusinski operators.

It was to develop a complex spectral theory that the present author introduced b-spaces and b-algebras [81]. The theorems he obtained can be applied without any loss of strength when topological algebras are involved. It is his experience that there are many instances where algebras with topologies can be replaced advantageously by algebras with bounded structures. The completeness hypothesis is useful in applications (a fundamental system of completant bounded sets exists), for many reasons. This is the reason why he lays the emphasis on b-spaces here and elsewhere.

Vector spaces with p-norming, or p-completant bounded structures can be introduced. It is not sure that their consideration is worth while unless some results can be obtained which use these bounded structure and cannot be obtained from the consideration of some, presumably finer completant bounded structure.

H. Buchwalter [12], G. Noël [60], [61], H. Hogbe-Nlend [40] have carried out research on the general linear and multilinear theory of spaces with convex bounded structures. Tensor products, spaces of mappings, the general relation between a topology and a bounded structure are considered in these theses. Noël is mainly concerned with an abelian category, which contains the category of b-spaces, and is involved in the author's "relative holomorphic functional calculus".

The reader can also be referred to an older set of lecture notes on the subject (ref. [63]) which has become recently available, and contain some of Buchwalter's and Hogbe-Nlend's results.

J. Sebastião e Silva [69] studied the Gâteaux and Fréchet differentials in topological vector spaces. He showed they involved some structures similar to those defined here.

Fischer [24], Cook and Fischer [17], Frölicher and Bucher [28] have studied systematically more general convergence structures than those considered here.

H. Hogbe-Nlend [41] is publishing a set of lecture notes about vector spaces with bounded structures. He did not have, like the present author, to wade through his contemporaries' skepticism, and considers vector spaces with general bounded structures as legitimate objects of research. Thanks to this difference in viewpoints, he solves quite a few problems which are not even considered here.

The theory is still in a state of flux. The author has published a minimal exposition of the theory [85], so that at least that much would be available to the general mathematical public. He also defined, considered differentiable function of class C_∞

with values in a b-space. F. Cnop-Grandsard [3] studied functions of class C_∞ with values in a b-space and polynomial growth at infinity. She showed that two definitions which are equivalent in the locally convex case are not equivalent when b-spaces are considered.

This author has once dared to ask a number of questions about convex bounded structures [82]. All the answers turned out to be negative ([83], [59], [20], etc.). It was a good lesson, a locally convex intuition is misleading when bounded structures are concerned. It is less difficult to prove theorems and find counter-examples than to put one's finger on the notions which will last once the theory has gone through its first sickness.

The topological vector spaces with the convexity property involved in proposition 3 are being studied by P. Turpin (Orsay) : It is possible to find a sequence λ_k of positive real numbers in such a way that, to each neighbourhood U of the origin there corresponds a neighbourhood V such that

$$U \supseteq \bigcup_N \Sigma_1^N \lambda_k V$$

His results are still unpublished.

The small boundedness is important in locally pseudo-convex space theory. It is also important in nuclear space theory.
When proving that a nuclear space can be embedded in s^A, T. Kōmura and Y. Kōmura [47] show really that the equicontinuous boundedness of the dual of a locally convex nuclear space is small.
A. Grothendieck ([37], chapter II, paragraph 2.4, proposition 9) had shown earlier that such a result was equivalent to the embedding theorem.

The definition of a nuclear b-space is straightforward ([40], or [100], def. 2.1 - chap. III. The proof of Kōmura and Kōmura's theorem shows easily that the boundedness of a nuclear b-space is small ([47], proposition 3).

G. Köthe [48] has already given an example where the limiting operations necessary to close off a subspace of a b-space are nasty. Köthe did not consider b-spaces then, of course, but considered sequential convergence in a locally convex space whose topological boundedness happens to be of type ($\mathfrak{S}$). Sequences converge for the boundedness in such space if they converge topologically.

A b-space if of type ($\mathfrak{S}$), or co-($\mathfrak{S}$) if a bounded completant $B' \supseteq B$ can be associated to each bounded B in such a way that B is relatively compact in $E_{B'}$.

Proposition 11, paragraph 5 gives examples of b-spaces with vanishing duals. Just take any $\mathfrak{F}$-space whose dual vanishes. The associated b-space of type $\mathfrak{F}$ will have a vanishing dual. A bounded linear form on the associated b-space of type $\mathfrak{F}$ must be topologically continuous, therefore identically zero.

Theorem 12 is due to Grothendieck ([38], chapter IV, paragraph 4, theorem 1) except for one detail. Grothendieck's statement only applies when some Hausdorff vector topology exists on E which is compatible with the boundedness.
There are examples of b-spaces of type $\mathcal{L}\mathfrak{F}$ with no such topology, but a b-space of type $\mathcal{L}\mathfrak{F}$ is a b-space. It is also sufficient to show that the graph is closed for the bounded structure.

A result equivalent with theorem 12 is also given by Hogbe-Nlend ([41], chapter VI, th. 2.1.). Hogbe's bounded structure is weaker than ours, but theorem 11 shows easily that the graph is closed for Hogbe's bounded structure if it is closed for ours so that the two results are equivalent.

M. De Wilde [18] had already shown that sequential closedness of the graph was in many cases, sufficient to prove a closed graph theorem. An analysis of De Wilde's proof shows that it is sufficient for the graph to be closed for the bounded structures. A further analysis of his proof, and a slight change of some definitions and of the statements of some theorems show that a closed graph theorem can be obtained for mappings $E \to F$ where E is an $\mathfrak{F}$-space, and F belongs to a class of b-spaces which contains b-spaces of type $\mathfrak{F}$, and is stable for countable (monic) direct limits and countable inverse limits. De Wilde considered only locally convex $\mathfrak{F}$-spaces and their countable direct and inverse limits.

CHAPTER III

Compactological spaces

1. Compactologies : Let E be a set. Let $\mathcal{B}$ be a bounded structure on E. For each $B \in \mathcal{B}$ let $\mathcal{T}(B)$ be a topology on B. Assume furthermore that :

a) $\mathcal{T}(B')$ induces $\mathcal{T}(B)$ on B when $B' \supseteq B$;
b) It is possible to associate $B' \in \mathcal{B}$ to every $B \in \mathcal{B}$ in such a way that $B' \supseteq B$ and $\mathcal{T}(B')$ is compact.

Definition 1 : When these conditions hold, $(E, \mathcal{B}, \mathcal{T})$ is a compactological space. If $(E_i, \mathcal{B}_i, \mathcal{T}_i)$, $i = 1, 2$ are compactological spaces, $u : E_1 \to E_2$ is a morphism if $uB \in \mathcal{B}_2$ for every $B \in \mathcal{B}_1$, $u : B \to uB$ being continuous for the topologies $\mathcal{T}_1(B)$, $\mathcal{T}_2(uB)$.

The following example of a compactology is standard. Let $(E, \mathcal{S})$ be a T_1 topology. Let $\mathcal{B}$ be the set of $B \subseteq E$ whose closure is compact (here as elsewhere, compact implies T_2, i.e. Hausdorff). For $B \in \mathcal{B}$, let $\mathcal{T}(B)$ be the topology induced by $\mathcal{S}$ on B. Then $(E, \mathcal{B}, \mathcal{T})$ is a compactological space.

This is easy. The only non trivial part of the proof is that of the fact that $B_1 \cup B_2$ is Hausdorff when B_1 and B_2 are closed and Hausdorff.
An ultrafilter $\mathcal{U}$ on $B_1 \cup B_2$ should not have more than one limit point in $B_1 \cup B_2$, but if e.g. $B_1 \in \mathcal{U}$, then $\mathcal{U}$ cannot converge to more than a point in B_1, because B_1 is Hausdorff ; $\mathcal{U}$ cannot converge to any point in $B_2 \setminus B_1$ because $B_2 \setminus B_1$ is open in $B_1 \cup B_2$.

We can generalize this example somewhat. Let $(E, \mathcal{T})$ be again a T_1 space. Let $\mathcal{B}$ be any boundedness on E, with a fundamental system of closed, compact, Hausdorff elements. For $B \in \mathcal{B}$, let $\mathcal{T}(B)$ be the topology induced on B by $\mathcal{S}$. Then $(E, \mathcal{B}, \mathcal{T})$ is again a compactology.

Proposition 1 : All compactologies can be obtained in this way.

This is obvious once we look at the direct limit topology $\mathcal{S}$ of a compactological space $(E, \mathcal{B}, \mathcal{T})$. A subset $X \subseteq E$ is closed for $\mathcal{S}$ if and only if $X \cap B$ is $\mathcal{T}(B)$-closed for every $B \in \mathcal{B}$. Then $\mathcal{S}$ is T_1, the compact elements of $\mathcal{B}$ are closed, and $\mathcal{S}$ induces on these compact elements the original topology $\mathcal{T}(B)$.

Remark : In general, $\mathcal{S}$ is not T_2. Let E be the space of polynomials which vanish at the origin. Let $\mathcal{B}$ be the set of $B \subseteq E$ which for some $\varepsilon > 0$ are relatively compact in E for uniform convergence on the interval $(-\varepsilon, \varepsilon)$. We topologize B by uniform convergence on the interval $(-\varepsilon, \varepsilon)$. It can then be shown that the open subsets of E are dense, when we topologize E by the direct limit topology. The compactology of E is a vector space compactology, but the direct limit topology $\mathcal{S}$ is not a vector space topology.

We must mention one more general nonesensical result about this direct limit topology. Let $F \subseteq E$ be closed for $\mathcal{S}$, then $F \cap B$ is closed for $\mathcal{T}(B)$ for all $B \in \mathcal{B}$, and $\mathcal{T}(F \cap B)$ is compact whenever $\mathcal{T}(B)$ is compact.
Let then $\mathcal{B}_1$ be the set of $B \cap F$ with $B \in \mathcal{B}$, and $\mathcal{T}_1$ be the restriction of $\mathcal{T}$ to $\mathcal{B}_1$. $(F, \mathcal{T}_1, \mathcal{B}_1)$ is in this way a compactological space. A direct limit topology $\mathcal{S}_1$ is defined in this way on F.

Proposition 2 : $\mathcal{S}_1$ is the topology induced by $\mathcal{S}$ on F.

This is trivial. Let $X \subseteq F$. Then X is $\mathcal{S}$-closed iff $X \cap B$ is $\mathcal{T}(B)$-closed for all $B \in \mathcal{B}$. It is $\mathcal{S}_1$-closed iff $X \cap (B \cap F)$ is $\mathcal{T}(B \cap F)$-closed for all $B \in \mathcal{B}$. The result follows since $X \cap B = X \cap (B \cap F)$ and $\mathcal{T}(B)$ induces $\mathcal{T}(B \cap F)$ on $B \cap F$.

We shall be considering compactologies on algebraic structures.

Definition 2.1 : Let $(G, \mathcal{B}, \mathcal{T})$ be a compactological space, with a group multiplication. Then $(G, \mathcal{B}, \mathcal{T})$ is a compactological group if $B \,.\, B'^{-1} \in \mathcal{B}$ when $B \in \mathcal{B}$, $B' \in \mathcal{B}$, the mapping $(x,y) \to x.y^{-1}$ $B \times B' \to B \,.\, B'^{-1}$ being continuous.

Definition 2.2 : Let $(\mathfrak{a}, \mathcal{B}, \mathcal{T})$ be a ring with a compactology. Then $(\mathfrak{a}, \mathcal{B}, \mathcal{T})$ is a compactological ring if $(\mathfrak{a}, \mathcal{B}, \mathcal{T})$ is additively a compactological group, and if furthermore $B_1 . B_2 \in \mathcal{B}$ when $B_1 \in \mathcal{B}$, $B_2 \in \mathcal{B}$, while $(x,y) \to x.y$ maps $B_1 \times B_2$ continuously into $B_1 . B_2$.

<u>Definition 2.3</u> : Let $(k, \mathcal{B}, \mathcal{T})$ be a field with a compactology. Then $(k, \mathcal{B}, \mathcal{T})$ is a compactological field if it is a compactological ring and $B^{-1} \in \mathcal{B}$ when $B \in \mathcal{B}$, is compact, and does not contain the origin, the mapping $x \to x^{-1}$, $B \to B^{-1}$ being continuous.

<u>Definition 2.4</u> : Let $(\mathfrak{a}, \mathcal{B}, \mathcal{T})$ be a compactological ring, and $(M, \mathcal{B}_1, \mathcal{T}_1)$ be an $\mathfrak{a}$-module with a compactology. Then $(M, \mathcal{B}_1, \mathcal{T}_1)$ is a compactological module if $B . B_1 \in \mathcal{B}_1$ when $B \in \mathcal{B}$, $B_1 \in \mathcal{B}_1$, the mapping $(a,m) \to a.m$ being continuous $B \times B_1 \to B . B_1$.

Let G be a topological group, $\mathfrak{a}$ a topological ring, k a topological field, or E a topological vector space. We find a group, a ring, a field, or a vector space compactology on $G, \mathfrak{a}, k$ or E respectively ; its elements are the relatively compact subsets. Their topology $\mathcal{T}(B)$ is the induced topology.

Not all group (or other) compactologies can be defined in this way. The example given above of a compactological space whose limit topology is not Hausdorff is a compactological group (even vector space) and no Hausdorff group topology is compatible with the given compactology.

We shall be considering real and complex compactological vector spaces. The compactology that we shall put on $\mathbb{R}$ or on $\mathbb{C}$ is the usual, topological compactology of the field.

2. <u>Countable compactologies</u> : A compactology is countable if there is a countable fundamental system of bounded sets. A countable compactological space is the union of an increasing sequence of compact spaces $(B_n, \mathcal{T}_n)$. The identity mappings $B_n \to B_{n+1}$ are continuous. A set B is bounded if $B \subseteq B_n$ for n large ; $\mathcal{T}(B)$ is the topology induced by $\mathcal{T}(B_n)$ on B.

We shall investigate the direct limit topology $\mathcal{S}$ of $E = \bigcup B_n$.

<u>Proposition 3</u> : A subset $X \subseteq E$ is compact for the topology $\mathcal{S}$ if, and only if $X \subseteq B_n$ for n large, and is closed.

The condition is trivially sufficient. Conversely, assume that X is not contained in any B_n. A sequence x_n of elements of X can then be found such that $x_n \notin B_n$. Let $Y = \{x_n\}$.
The set $Y \cap B_n$ is then finite for all k, it is thus closed and discrete. Y is therefore closed, is discrete for the induced topology, and is infinite. But $Y \subseteq X$, thus X cannot be compact.

This proposition says that the compactology of a space with countable compactology can be recovered when the topology $\mathcal{S}$ is given.

<u>Lemma 1</u> : Let $E = \bigcup B_n$ be a countable compactological space.
Let $A_n \subseteq B_n$ for all n and assume that A_{n+1} is a neighbourhood of A_n in the topological space B_n. Then $A = \bigcup A_n$ is open in E for the limit topology.

This is well known, $\bigcup A_n = \bigcup \mathring{A}_n$ (where $\mathring{A}_n$ is the interior of A_n in the topological space B_n), and

$$(\bigcup A_n) \cap B_k = \bigcup_{n > k} (\mathring{A}_n \cap B_k)$$

is open in B_k for all k.

<u>Proposition 4</u> : The direct limit topology of a countable compactological space is normal.

We assume $E = \bigcup B_n$ is a countable compactological space. We let F, G be disjoint closed sets. We assume that disjoint closed neighbourhoods U_{n-1}, V_{n-1} of $F \cap B_{n-1}$, $G \cap B_{n-1}$ have been found. The sets $U_{n-1} \cup (F \cap B_{n-1})$ and $V_{n-1} \cup (G \cap B_n)$ are closed, and disjoint.
Then U_n, V_n will be disjoint, closed neighbourhoods of these two sets. The induction is started by assuming $B_o = \emptyset$, so that the initial statement is empty.

The construction yields two disjoint, open sets $\bigcup U_n$ and $\bigcup V_n$ which contain F and G respectively.

Let now $E = \bigcup B_n$ and $F = \bigcup C_n$ be two countable compactological spaces. $E \times F = \bigcup (B_n \times C_n)$ is in a natural way a countable compactological space. We have a priori two topologies on $E \times F$, a product topology and a limit topology.

Proposition 5 : These two topologies coincide.

It is trivial that the product topology of $E \times F$ is weaker than the limit topology. We must show it is stronger. Let $(x,y) \in E \times F$ and U be an open neighbourhood of (x,y) for the limit topology.

We assume without loss of generality that $x \in B_1$, $y \in C_1$. Assume further that closed sets A_{n-1}, A'_{n-1} have been constructed, and are contained in B_{n-1}, C_{n-1} respectively, while $A_{n-1} \times A'_{n-1} \subseteq U$. We shall choose closed neighbourhoods A_n, A'_n of A_{n-1}, A'_{n-1} in B_n, C_n respectively, in such a way that $A_n \times A'_n \subseteq U$. Then $\bigcup A_n = A$, $\bigcup A'_n = A'$ are open sets and $A \times A' \subseteq U$, so that U is a neighbourhood of (x,y) for the product topology. To start the induction, we let $A_o = \{x\}$, $A'_o = \{y\}$.

3. Countable compactologies on groups, rings and modules

Proposition 6.1 : Let G or $\mathfrak{a}$ be a group or a ring with a countable group or ring compactology. The direct limit topology of G or $\mathfrak{a}$ is a group or a ring topology respectively.

Proposition 6.2 : Let $\mathfrak{a}$ be a ring with a countable ring compactology and M an $\mathfrak{a}$-module with a countable $\mathfrak{a}$-module compactology. Then M is a topological module on the topological ring $\mathfrak{a}$ when we consider on $\mathfrak{a}$ and on M the respective direct limit topologies.

These results are obvious. In the group case, we must check that $(x,y) \to x^{-1} y$ maps $G \times G$ continuously into G. Since the topology of $G \times G$ is the direct limit of the topologies $B \times B'$, B and B' compact in G (proposition 5) it is sufficient to verify that the restriction of this mapping to $B \times B'$ is continuous. And this is part of the hypothesis.

The ring and the module case are handled in a similar way.

<u>Proposition 7</u> : The direct limit topology of a group G with a countable compactology is sequentially complete.

This will show that a ring $\mathfrak{a}$ with a countable compactology or a module M with a countable compactology is sequentially complete with the limit topology, since completeness of a ring or a module is equivalent with completeness of the additive group.

We assume that $G = \bigcup X_n$, where each X_n is compact, and $X_n^{-1} \cdot X_n \subseteq X_{n+1}$. Let $\{g_k\}$ be a Cauchy sequence, $g_{k'}^{-1} \cdot g_k \to e$ (the unit) when k, k' tend to infinity.

Let $\nu_1, \nu_2, \ldots$ be a sequence of positive integers. The set

$$\{g_{k'}^{-1} \; g_k \mid k \leq k' \leq \nu_k\}$$

is a countable set, which once ordered becomes a sequence tending to e, so the set is relatively compact, is contained in X_n for large n.
We let

$$n(\nu) = \inf \{n \mid \# (\{g_{k'}^{-1} \cdot g_k \mid k \leq k' \leq \nu_k\} \setminus X_n) < \infty$$

i.e. $n(\nu)$ is the smallest value of n for which the $g_{k'}^{-1} \; g_k$ with $k \leq k' \leq \nu_k$ all belong to X_n, except for a finite number of possible exceptions. Then $n(\nu)$ is an increasing function on $\mathbb{N}^{\mathbb{N}}$. It is constant on the equivalence classes for the relation

$$(\nu \equiv \nu') \iff (\# \{k \mid \nu_k \neq \nu'_k\} < \infty)$$

Cantor's diagonal shows that such a function is bounded.

This gives an N, independent of the sequence ν, and such that the number of (k,k') such that $g_{k'}^{-1} \; g_k \notin X_N$ with $k \leq k' \leq \nu_k$ is always finite. And this can only happen if $g_{k'}^{-1} \; g_k$ $(k \leq k')$ always belongs to X_N, except for a finite number of exceptional values of k. So our Cauchy sequence is in a compact set, and converges.

<u>Proposition 8.1</u> : Let G be a compactological group. Let $\{B_n\}$ be a countable system of compact subsets of G. The strongest group

compactology on the group generated by $\bigcup B_n$, and for which the sets B_n are bounded, and have their given topologies, is a countable one.

Proposition 8.2 : Let $\mathfrak{a}$ be a compactological ring. Let $\{B_n\}$ be a countable system of compact subsets of $\mathfrak{a}$. The strongest ring compactology on the ring generated by $\bigcup B_n$, and for which the sets B_n are bounded, and have their given topologies, is a countable one.

Proposition 8.3 : Let M be a compactological module over a ring $\mathfrak{a}$ with a countable compactology. Let $\{B_n\}$ be a countable set of compact subsets of M. The strongest module compactology on the module generated by $\bigcup B_n$, for which the sets B_n are bounded and have their given topology, is a countable compactology.

We shall speak of the group, of the ring, or of the module generated by the sets $\{B_n\}$.

The results are rather trivial. In the group case, we would put

$$C_n = B_n \cup B_n^{-1} \cup C_{n-1}^{-1} \cdot C_{n-1}$$

In the ring case, we would put

$$C_n = B_n \cup (-B_n) \cup (C_{n-1} - C_{n-1}) \cup C_{n-1} \cdot C_{n-1}$$

In the module case, we would consider an increasing fundamental system $\{B'_k\}$ of compact subsets of $\mathfrak{a}$, and put

$$C_n = B_n \cup (-B_n) \cup (C_{n-1} - C_{n-1}) \cup B'_k \, (C_{n-1} - C_{n-1})$$

In all three cases, we would begin the induction letting C_o be the unit, or the additive neutral element. The sets C_n are a fundamental system of bounded sets for the compactology we consider on the generated group, ring, or module.

Vector spaces are special modules.

4. Fields with countable compactologies : The situation is somewhat more complicated when fields are considered.

Proposition 9 : The compact subsets of a field with a countable compactology are metrizable.

Let B be such a compact set. The "entourages" of the uniform structure of B are the inverse images of the neighbourhoods of the origin in $B - B$ for the mapping $(x,y) \to y - x$. The compact set is metrizable if the origin has a countable fundamental set of neighbourhoods in $B - B$.

So it is sufficient to show that the origin has a countable fundamental system of neighbourhoods in B, if B is compact, $0 \in B$. Let $\{B_n\}$ be a fundamental sequence of compact sets of our compactological field, let $C_n = \{x \in B \mid x \neq 0, x^{-1} \in B_n\}$. Then C_n is closed, does not contain the origin, hence $B \setminus C_n$ is a neighbourhood of the origin in B. If U is a neighbourhood of the origin in B, take U open, then $\{x^{-1} \mid x \in B \setminus U\}$ is compact, contained in B_n for some n, i.e. $U \supseteq B \setminus C_n$ for some n.

We can now state the analogue of proposition 8.

Proposition 10 : Let k be a compactological field. Let $\{B_n\}$ be a sequence of compact metrizable subsets of k. Let k_1 be the field generated by $\bigcup B_n$. The strongest compactology on k_1, for which the sets B_n are bounded, and have their given topologies, is a countable compactology.

The construction of the compact subsets C_n of k_1 is somewhat more complicated than in proposition 8. We let $C_o = \{0\}$ and assume that C_n has been constructed, is compact and metrizable.
We also assume that a fundamental sequence of open neighbourhoods of the origin $U_{n.k}$ of the origin in C_n is given.
We put $C_{n.k} = C_n \setminus U_{n.k}$.
We go on in the induction, putting

$$C_{n+1} = B_{n+1} \cup (C_n - C_n) \cup (C_n \cdot C_n) \cup C_{n.n}^{-1}$$

This is a metrizable compact set. It contains C_n. Its topology induces the given topology on C_n. We can choose the open neighbourhoods $U_{n+1.k}$ of the origin in C_{n+1} in such a way that
$U_{n+1.k} \cap C_n \subseteq U_{n.k}$, i.e. $C_{n+1.k} \supseteq C_{n.k}$.

It is now clear, or easy to verify that $\bigcup C_n$ is the field generated by $\bigcup B_n$, that the compactology determined by $\{C_n\}$ is a field compactology, and is stronger than any field compactology for which the sets B_n are compact and have their given topologies.

<u>Proposition 11</u> : The direct limit topology of a field with a countable compactology is a field topology.

Let k be the field, $\{B_n\}$ be a fundamental sequence of compact subsets of k. We already know that k is a topological ring with its direct limit topology. We must still show that $x \to x^{-1}$ is continuous. It will be sufficient to verify continuity on some neighbourhood of the unit.

Let A be a closed neighbourhood of the unit. Then $A = \bigcup A_n$ where $A_n = A \cap B_n$; the induced topology of A is the direct limit of the topologies of the A_n. It is therefore sufficient to show that $x \to x^{-1}$ maps A_n continuously into k, but A_n is compact, therefore mapped continuously into $A_n^{-1} \subsetneq k$.

5. <u>An example</u> : We have assumed that some group compactology existed, on a group G, for which the given compact sets happened to be compact. The following gives an inkling of the problems encountered when we do not presuppose the existence of such a compactology.

<u>Proposition 12</u> : Let G be a group. Let X be a symmetric, generating subset of G, and τ be a compact topology on X. A group topology τ_1 can be found on G which induces τ on X iff the sets

$$\{(x_1, \ldots, x_n) \in X \times \ldots \times X \mid x_1 \ldots x_n = e\}$$

are all closed.

This is easy. Half is trivial, if a group topology τ_1 exists with the required properties, then these sets are of course closed.

If these sets are closed, we begin by verifying that inversion $x \to x^{-1}$ maps X continuously into itself. But the graph of the mapping is

$$\{(x,y) \in X \times X \mid x.y = e\}$$

This is compact in $X \times X$ and the mappings $(x,y) \to x$, $(x,y) \to y$ are homeomorphisms.

We then observe that the relation

$$x_1 \ldots x_n = x'_1 \ldots x'_n$$

is a closed equivalence on $X \times \ldots \times X$. The quotient can be identified with X^n, and the identity $X^n \to X^{n+1}$ is continuous. We obtain in this way a countable compactology on $G = \bigcup X^n$, the required topology is the direct limit topology of this compactological group. The hypothesis that the set of $(x_1, \ldots, x_n)$ such that $x_1 \ldots x_n = e$ is closed for all n cannot be weakened essentially :

<u>Proposition 13</u> : For every $n > 2$, it is possible to find a group G, a subset $X \subseteq G$, and a compact topology τ on X such that

$$\{(x_1, \ldots, x_k) \in X \times \ldots \times X \mid x_1 \ldots x_k = e\}$$

is closed for $k < n$, but is not closed when $k = n$.

X will be a compact space, G_1 will be the free commutative group generated by X, H will be the normal subgroup of G_1 generated by $\{t^n \mid t \in U\}$ where U is a non-closed subset of X, and $G = G_1/H$. The structure of G is well known, we can embed X in G if $n > 1$.

We let $Y = X \cup X^{-1}$ so Y is symmetric and topologize Y in an obvious way : X and X^{-1} are disjoint in G if $n > 2$. And clearly Y is the desired example.

6. <u>Countable convex compactologies</u>

<u>Definition 3</u> : (B,τ) is a compact disc in the real or complex vector space E if B is absolutely convex, if τ is a compact topology on B, if the mapping $(x,y) \to \frac{x+y}{2}$, $B \times B \to B$ is continuous, and if the origin in B has a fundamental system of absolutely convex neighbourhoods for τ.

Definition 4 : Let $(E, \mathcal{B}, \mathcal{T})$ be a compactological vector space, such that every $B \in \mathcal{B}$ is contained in some $B_1 \in \mathcal{B}$ such that $(B_1, \mathcal{T}(B_1))$ is a compact disc. Then $(E, \mathcal{B}, \mathcal{T})$ is a convex compactological space.

We shall say that B_1, is a compact disc of the compactology if $B_1 \in \mathcal{B}$ and $(B_1, \mathcal{T}(B_1))$ is a compact disc.

Proposition 14 : Let $(E, \mathcal{B}, \mathcal{T})$ be a compactological vector space, whose compactology is countable and convex. The direct limit topology $\mathcal{J}$ of E is then locally convex.

Let $(B, \mathcal{T})$ be a compact disc. Associate to each neighbourhood U of the origin the set

$$U_1 = \{(x,y) \in B \times B \mid x - y \in 2U\}$$

This is the inverse image of U by the mapping $(x,y) \to (x - y)/2$. It is therefore a neighbourhood of the diagonal in $B \times B$. When U ranges over the filter of neighbourhoods of the origin in $\mathcal{B}$, U_1 ranges over a filter basis, with a closed sub-basis, each U_1 is a neighbourhood of the diagonal, the intersection of the sets U_1 is the diagonal. In other words, this filter on $B \times B$ is the filter of all entourages for the compact uniformity of B. This shows that the uniformity is convex.

We need this fact to prove that a closed convex subset of B has a fundamental system of convex neighbourhoods. These neighbourhoods are uniform, and a fundamental system of uniform convex neighbourhoods can be found.

Proposition 14 now follows by a straightforward application of lemma 1. Let U be an open neighbourhood of the origin. Let $A_o = \{0\}$ assume $A_n \subseteq U$ has been chosen, is compact and absolutely convex in B_n. Choose a compact, absolutely convex neighbourhood A_{n+1} of A_n in B_{n+1}, such that $A_{n+1} \subseteq U$. The union $\bigcup_n A_n$ will be the required convex neighbourhood of the origin.

Let now $(E, \mathcal{B}, \mathcal{T})$ be a compactological vector space, with a countable convex compactology. Let $F \subseteq E$, assume F closed for the direct limit topology, then $(\mathcal{B}, \mathcal{T})$ induces a compactology on

F, whose elements are the $B \cap F$, $B \in \mathcal{B}$, the topology of $B \cap F$ being $\mathcal{T}(B \cap F)$: this is compact if B is compact. The direct limit topology of E induces on F the direct limit topology of F. So we have the following result.

<u>Proposition 15</u> : A linear form u on F whose restriction to each $B \cap F$, $B \in \mathcal{B}$ is continuous for $\mathcal{T}(B \cap F)$ extends to a linear u on E with continuous restriction to each $B \in \mathcal{B}$.

Just apply the Hahn-Banach theorem.

The situation is not as simple when F is not closed. For each compact absolutely convex $B \in \mathcal{B}$, consider $B \cap F$, then its closure $\overline{B \cap F}$, which does not really depend on B, but on $B \cap F$. A linear form on F is uniformly continuous on $B \cap F$ if it is continuous at the origin (this is a famous lemma of Grothendieck). This linear form extends therefore to $\overline{B \cap F}$.

We shall let $\lim F = \bigcup_{B \in \mathcal{B}} \overline{B \cap F}$ - the notation is reasonably consistent with the one that we adopted in Definition 2.18. Just as in the b-space case, $\lim F$ may be or not be a closed subspace of E.

<u>Proposition 16</u> : If $\lim F$ is a closed subspace of E, every linear form on F with continuous restriction to $B \cap F$ for every F has an extension, which is linear on E, and has a continuous restriction to every $B \in \mathcal{B}$. Conversely, if $\lim F$ is not a closed subspace of E, it is possible to find a linear form on F with a continuous restriction to every $B \cap F$, which does not have any extension to E, continuous on every $B \in \mathcal{B}$ for the topology $\mathcal{T}(B)$.

Applying proposition 15, and the remarks above, we find an extension of u_o if u_o is defined and has the required continuity properties on F, when $\lim F$ is closed.

Assume now $F_1 = \lim F$ is not closed. Consider the linear forms on F_1 whose restrictions to $B \cap F$ is $\mathcal{T}(\overline{B \cap F})$ continuous. Topologized uniform convergence on the sets $\overline{B \cap F}$, this space B of linear forms is a locally convex $\mathcal{F}$-space.

A simple application of the bipolar theorem (see also Proposition 17) shows that G^* is isomorphic to F_1, this isomorphism identifying the equicontinuous boundedness of G^* with the boundedness generated on F_1 by the sets $\overline{B \cap F}$. Since G is barreled, a pointwise bounded subset of F_1 is equicontinuous. In other words, we can find $g \in G$ which is unbounded on X if $X \subseteq F_1$ is not contained in any $\overline{B \cap F}$.

We assume that F_1 is not closed. Some $B_o \in \mathcal{B}$ can then be found such that $B_o \cap F_1$ is not contained in any $\overline{B \cap F}$. Some $g \in G$ can be found which is unbounded on $B_o \cap F_1$. The restriction of g to F is the required linear form. If it had an extension of the type we are looking for, the extension would coincide with g on F_1 and be bounded on B_o, but g is unbounded on $B_o \cap F$.

7. <u>Convex compactologies and duality</u> : We shall describe two "duality" functors from the category of convex compactological spaces to that of locally convex spaces and conversely.

<u>Definition 5</u> : Let $(F, \mathcal{T})$ be a locally convex space. Then $Du_1(F, \mathcal{T})$, or $Du_1 F$ when the topology is well defined, is the space of continuous linear forms on F, with the equicontinuous boundedness, each bounded set being equipped with its weak topology.

<u>Definition 6</u> : Let $(E, \mathcal{B}, \mathcal{T})$ be a convex compactological space. Then $Du_2(E, \mathcal{B}, \mathcal{T})$ is the vector space of linear forms on E, whose restriction to each $B \in \mathcal{B}$ is continuous for the topology $\mathcal{T}(B)$, topologized by uniform convergence on the elements of $\mathcal{B}$.

The completion functor in the category of locally convex spaces is well known. The completion of F will be called $\hat{F}$. We have $Du_1 F = Du_2 \hat{F}$ of course (the two spaces are naturally isomorphic).

We shall need a "separated quotient" functor on the category of convex compactological spaces. Let $(E, \mathcal{B}, \mathcal{T})$ be a convex compactological space. Let E_o be $\bigcap \operatorname{Ker} u$ where u ranges over $Du_2(E, \mathcal{B}, \mathcal{T})$. Let $E_1 = E/E_o$, for $B \in \mathcal{B}$, let $B_1 = qB$, where $q : E \to E_1$ is the quotient mapping. Let $\mathcal{T}_1(B_1)$ be the quotient topology on B_1 ; this is a Hausdorff topology because the $u \in Du_2 E$

separate E_1, so it is a compact topology when B is compact, and clearly $(B_1, \mathcal{T}_1)$ is a compact disc when $(B, \mathcal{T})$ is one.

Definition 7 : Let $(E, \mathcal{B}, \mathcal{T})$ be a convex compactological space. Sep $(E, \mathcal{B}, \mathcal{T})$ will be $E_1 = E/E_o$, with the compactology defined by the qB, $B \in \mathcal{B}$, each topologized by the quotient topology, where $E_o = \bigcap \mathrm{Ker}\, u$, $u \in Du_2 E$, and $q : E \to E_1 = E/E_o$ is the quotient mapping.

Proposition 17 : $Du_2 \circ Du_1$ is the completion functor, whereas $Du_1 \circ Du_2 = \mathrm{Sep}$.

Since $Du_2 = Du_2 \circ \mathrm{Sep}$, and Sep is an idempotent functor, both modulo isomorphism of functors, it is sufficient to prove that $Du_2 \circ Du_1$ is the identity on convex compactological spaces $(E, \mathcal{B}, \mathcal{T})$ which are separated by $Du_2(E, \mathcal{B}, \mathcal{T}) = F$.
And this follows from the Mackey-Arens theorem, applied to the dual pair (E,F). (The Mackey-Arens theorem is itself a simple application of the bipolar theorem).

The fact that $Du_2 \circ Du_1$ is the completion functor is the Grothendieck completion theorem.

Notes and remarks : The compactological spaces the author met first were the convex ones. They are useful in many Functional Analytic constructions involving duality. He regrets that he is unable to include a survey of such constructions in these lecture notes, but time is too short. The definition of convex compactological spaces being far from obvious, the author called them "farfelu" initially, far-fetched might be a free English translation.

The theorems contained in paragraphs 6 and 7 are complicated rewordings of the Banach-Dieudonné theorem, of the Grothendieck completion theorem, and of the Mackey-Arens theorem. A few simple applications of these theorems are included.

Some b-spaces have an obvious unique compactology. They are the "infra-$\mathcal{S}$ spaces". It is possible to associate to each bounded subset B of the space E a bounded completant $B' \supseteq B$ in such a way that B is weakly relatively compact in $E_{B'}$. If B is relatively compact in $E_{B'}$, for a suitable choice of B', then E is of type $(\mathcal{S})$.

Infra-$\mathcal{S}$ spaces with countable bounded structures are countable compactological spaces. We may apply proposition 16 to such spaces. The extension theorem obtained in this way is due to H. Hogbe-Nlend [41].
C. Foias and G. Marinescu [27] had proved a special case of this theorem, modulo a slight misstatement, corrected by Słowikowski [72]. The statement given here is in fact analogous to an earlier statement of the same theorem due to the author [85].

A b-space with a countable boundedness of type $(\mathcal{S})$ is called a "Silva" space, after J. Sebastião e Silva who introduced these spaces and showed their importance in applications [68].

It is much later that the author noticed that other compactological spaces could be useful - in 1968 in fact, during a stay at the Mathematics Department of the University of Warwick. The results given in paragraphs 1 trough 4 are mimeographed by the Bordeaux Functional Analysis Seminar [88], [89].

The results given in paragraphs 1 and 2 would probably not be considered new by any General Topologist. Those of paragraphs 3, 4 and 5 would not interest the General Topologist. The author hopes that the Functional Analyst will deem them interesting when he gets through reading these lecture notes.

The fact that Silva spaces, and spaces of type $(\mathcal{S})$ have a reasonable generalization can still be mentioned. Let $(E, \mathfrak{B}, \mathfrak{T})$ be a compactological space. Assume that it is possible to associate to every $B \in \mathfrak{B}$ some $B' \in \mathfrak{B}$ in such a way that to every $\varepsilon > 0$ a finite family $\{x_1, \ldots, x_k\}$ of elements of B can be found with $B \subseteq \bigcup_1^k (x_i + \varepsilon B')$. The topology $\mathfrak{T}(B)$ can then be recovered from the bounded structure $\mathfrak{B}$ in a way not dissimilar from the way the bounded structure of a Silva space gives a family of topologies on the bounded sets.

Compactological spaces have been studied by H. Buchwalter in his thesis [13].

CHAPTER IV

Differentiable vector-valued functions

We consider functions of class C_r or C_∞ on a manifold U with values in a non locally convex topological vector space. These functions spaces have been defined by P. Turpin and the author [8]. The definition given here is not the original one, but is equivalent to it (cf. proposition 3).

The reader will notice our definition is related to H. Whitney's definition of a function of class C_r on a closed subset of $\mathbb{R}^n$ ([91], or [30]). This is not surprising.

Vector valued functions taking their values in a general topological vector space can be defined on closed sets. An extension theorem, similar to Whitney's can be proved. If X is a closed subset of U, if $F_r(U)$ is defined as in paragraph 1, if $F_r(X)$ is the space of elements of $F_r(U)$ whose support is in X, Whitney's extension theorem shows that $F_r(X)$ is a complemented subspace of $F_r(U)$. The proof of this assertion will be left to the reader.

1. The space : $F_r(U)$. Let U be a manifold which is countable at infinity. $\mathcal{E}(U)$ is the space of functions of class C_∞ on U. If r is a non negative integer, $\mathcal{E}_r(U)$ is the space of functions of class C_r on U. If $r \in \mathbb{R}_+$ is not an integer, $\mathcal{E}_r(U)$ is the space of functions of class $C_{[r]}$ on U whose derivatives of order $[r]$ satisfy a o-Hölder condition of exponent $r - [r]$.

We put on these spaces the obvious, Fréchet topology : uniform convergence on compact sets of the functions and of its derivatives up to the order $[r]$, along with (when r is not an integer) uniform convergence on compact subsets of V x V of expressions such as

$$|x - y|^{[r]-r} \left(\frac{\partial^{[r]} f}{\partial x^{[r]}} (x) - \frac{\partial^{[r]} f}{\partial x^{[r]}} (y) \right)$$

where $\partial^{[r]} f/\partial x^{[r]}$ stands for any derivative of order $[r]$ of f, with respect to the coordinate system of V, while V ranges over all (or enough) coordinate neighbourhoods. We note that the o-Hölder condition means that the expression whose convergence we must prove is a continuous function on $V \times V$ which vanishes on the diagonal.

$\mathcal{E}^*(U)$, and $\mathcal{E}_r^*(U)$ are the duals of $\mathcal{E}(U)$, or $\mathcal{E}_r(U)$ respectively. These are convex compactological spaces, with countable compactologies.

Let now $\varphi(u)$ be the number of elements of the support of an element of $\mathcal{E}^*(U)$. We observe that φ is lower semi-continuous for the weak-star topology. Also $\varphi(u + v) \leq \varphi(u) + \varphi(v)$, if s is scalar, $s \neq 0$ then $\varphi(su) = \varphi(u)$.

<u>Definition 1</u> : $F_r(U)$, or respectively $F(U)$ is the space of elements with finite support of $\mathcal{E}_r^*(U)$ or $\mathcal{E}^*(U)$. A subset B of $F_r(U)$ or of $F(U)$ is bounded if it is bounded in $\mathcal{E}_r^*(U)$ or $\mathcal{E}^*(U)$, and if φ is a bounded function on B. We topologize B with the weak topology $\sigma(\mathcal{E}^*(U), \mathcal{E}(U))$.

Clearly, $F_r(U)$ and $F(U)$ are countable compactological spaces. We shall also consider the evaluation mapping, $i: U \to F_r(U)$ defined when $x \in U$, $f \in \mathcal{E}_r(U)$, by $\langle ix, f\rangle = f(x)$.

<u>Definition 2</u> : Let E be a topological vector space, and U a manifold. A mapping $a: U \to E$ is of class C_r if it is possible to find $a_1: F_r(U) \to E$, which is linear, has a continuous restriction to the bounded subsets of $F_r(U)$, and is such that $a = a_1 \circ i$. This function space, $C_r(U,E)$ is topologized by uniform convergence of a_1, on bounded subsets of $F_r(U)$.

<u>Definition 2'</u> : $C_\infty(U,E)$ is the space of mappings $a: U \to E$ which have an extension $a_1: F(U) \to E$ which is linear and has a continuous restriction to the bounded subsets of $F(U)$.

The spaces $C_r(U,E)$, $C_\infty(U,E)$ are topological vector spaces. They are complete if E is complete, metrizable if E is metrizable. It is clear that each element of $F_r(U)$, or $F(U)$, is a limit of linear combinations of elements ix, and that the extension a_1 is determined by $a = a_1 \circ i$.

<u>Proposition 1</u> : Let $r < r'$, and let B be a compact subset of $F_r(U)$. A bounded subset B' of $F_{r'}(U)$, and a continuous distance $d(b, b')$ on B can be found in such a way that $b_1 - b_2 \in d(b_1, b_2)B'$ when $b_1, b_2 \in B$.

B is bounded in $F_r(U)$, and is therefore equicontinuous in $\mathcal{E}^*_r(U)$. The supports of the elements of B are contained in some compact subset K of U. We let V be open, relatively compact in U, V containing K.

The restriction mapping $\mathcal{E}_r(U) \to \mathcal{E}_{r'}(V)$ is compact. It is possible to find an absolutely convex, equicontinuous $B_1 \subseteq \mathcal{E}^*_r(V)$ in such a way that the Minkowski functional ν_{B_1} of B_1 induces on B the given topology $\mathcal{T}(B)$. We put $d(b_1, b_2) = \nu_{B_1}(b_1 - b_2)$ and let

$$B' = \left\{ \frac{b - b'}{d(b, b')} \;\middle|\; (b, b') \in B \times B \right\}$$

This is contained in B_1, therefore equicontinuous in $\mathcal{E}^*_{r'}(V)$, therefore a fortiori in $\mathcal{E}^*_{r'}(U)$. If the supports of the elements of B have at most k elements, those of the elements of B' have at most $2k$ elements, so B' is bounded in $F_{r'}(U)$. This proves proposition 1.

We let F_{r-o} be the union of the spaces $F_{r'}$, $r' < r$, a subset of F_{r-o} being bounded if it is bounded in $F_{r'}$ for some $r' < r$.

<u>Corollary 1</u> : The spaces $F_{r-o}(U)$, $F_\infty(U)$ are (non convex) compactological spaces of "Silva" type.

They are of countable type, obviously. Also, for every bounded B, one can find a bounded $B' \supseteq B$, and a continuous distance $d(b_1, b_2)$ on B in such a way that $b_1 - b_2 \in d(b_1, b_2)B'$ for all b_1, b_2 in B.

<u>Corollary 2</u> : A mapping $u : U \to E$ is of class C_r if $u : u_1 \circ i$ where $u_1 : F_{r'}(U) \to E$ is a bounded linear mapping, and $r' > r$.

Corollary 2' : A mapping $u: U \to E$ is of class C_∞ (respectively of class C_{r-o}) if $u = u_1 \circ i$ where $u_1 : F(U) \to E$ (respectively $u_1 : F_{r-o} \to E$) is a bounded linear mapping.

Continuity of u_1 on B follows from the boundedness of u_1 on B', where B' is suitably chosen.

2. Generation of $F_r(U)$. Let V be a coordinate neighbourhood in U, and $p = (p_1, \ldots, p_n) \in \mathbb{N}^n$ be such that $|p| = p_1 + \ldots + p_n \leq r$. Let then

$$i^{(p)}(x) = \frac{\partial^{|p|} i(x)}{\partial x^p}$$

and

$$T_{r-|p|}\, i^{(p)}(x) = \sum_{|q| \leq r-|p|} i^{(p+q)}(x) \frac{(y-x)^q}{q!}$$

Thus $i^{(p)}$ is the derivative of order $p = (p_1, \ldots, p_n)$ of the mapping $x \to i(x)$, and $T_{r-|p|}\, i^{(p)}(x,y)$ is the value in y of the Taylor expansion of $i^{(p)}$ around x.

Let now K be a compact subset of V. The following sets are relatively compact in $F_r(U)$:

$$X_p(K) = \{i^{(p)}(x) \mid x \in K\}$$

$$Y_{r.p}(K) = \left\{ |y-x|^{|p|-r} \left[i^{(p)}(y) - T_{r-|p|}(x,y) \right] \,\middle|\, (x,y) \in K \times K \right\}$$

As a matter of fact, X_p is compact, and so is $Y_{r.p} \cup \{0\}$.

Proposition 2 : The compactology of $F_r(U)$ is generated by the sets $X_p(K)$, $Y_{r.p}(K) \cup \{0\}$, where $p \in \mathbb{N}^n$ takes on all values such that $|p| \leq r$, and where K ranges over a system of compact sets, each one contained in a coordinate neighbourhood, the interiors of the compact sets considered covering U.

The notations $X_p(K)$, $Y_{r.p}(K)$ are not quite satisfactory since these sets depend on the coordinate system considered on K. The objection is not serious, especially since it is sufficient to prove proposition 2 when U is a single coordinate neighbourhood, i.e. when U is open in $\mathbb{R}^n$, and it is then sufficient to consider a single coordinate system.

Let B be a bounded subset of $F_r(U)$. The elements of B have their supports in a compact subset K of U. These supports are finite sets, with at most k elements, where k is an integer depending on B. We shall prove proposition 2 by induction on k.

If $k = 0$, $B = \{0\}$ and belongs to the compactology generated by the sets $X_p(K)$, $Y_{r.p}(K)$. This is sufficient to initiate the induction.

The proof will be easier to understand however if we consider further, small values of k.

If $k = 1$, the elements of B are distributions with single point supports. For each $a \in B$, we have a point x_a, and an expression

$$a = \sum_{|p| \leq r} c_p(a) \frac{\partial^{p_i}}{\partial x^p} (x_a)$$

The set of polynomials $(x - x_a)^p/p!$ is bounded in $\mathcal{E}_r(U)$,

$$c_p(a) = \langle a, \frac{(x - x_a)^p}{p!} \rangle$$

so $c_p(a)$ is a bounded function of a, and putting together all the information that we have, we see that B is contained in a finite linear combination of sets such as $X_p(K)$.

If $k = 2$, we may split up $B = B_1 \cup B_2$ where the supports of the elements of B_1 have at most one element, while the supports of the elements of B_2 have two elements. We have already disposed of B_1.

If $a \in B_2$, the points of the support of a will be called x_a and y_a. We let $d_a = |x_a - y_a|$ and split $B_2 = B'_2 \cup B''_2$ where $a \in B'_2$ if $d_a \geq 1$, while $a \in B''_2$ if $d_a < 1$.
We choose a function $\varphi(x)$, equal to one on a neighbourhood of the origin, and with compact support in the unit ball.

When $a \in B'_2$, we let $a_1 = \varphi(x - x_a).a$. This has one point support (at x_a) and is bounded independently of a, while $a - a_1$ also has one point support and is also bounded independently of a. The considerations developed when $k = 1$ apply both to a_1, and to $a - a_1$, B'_2 belongs to the boundedness generated by the sets $X_p(K)$.

When $a \in B''_2$, we put $a_1 = \varphi((x - x_a)/d_a).a$. Then a_1 again has its support in x_a ; it is not bounded independently of a. If

$$a_1 = \sum c_{p.a}\, i^{(p)}(x_a)$$

then $d_a^{|p|-r} c_{pa}$ is bounded independently of a however, because

$$d^{|p|-r} \varphi((x - x_o)/d)\,(x - x_o)^p$$

is bounded in $\mathcal{E}_r(U)$ independently of $d \in \mathbb{R}_+$, $x_o \in \mathbb{R}^n$, and

$$c_{p.a} = \langle a, \varphi(\frac{x - x_a}{d_a})\,(x - x_a)^p\rangle$$

Let now

$$a'_1 = \sum c_{p.a}\,(i^{(p)}(x_a) - T_{r-|p|}\, i^{(p)}(y_a, x_a))$$

We see that a'_1 is a linear combination of elements of the sets $Y_{r.p}(K)$. The number of terms is finite, the coefficients are bounded. So a'_1 ranges over an element of the boundedness generated by the $Y_{r.p}(K)$ when a ranges over B''_2. And $a - a'_1$ has one point support and is bounded. This disposes of B''_2 when $k = 2$.

All the essential elements of the induction proof are now in evidence.

Let B be bounded in $F_r(U)$, assume that $\varphi(a) \leq k$ when $a \in B$. Then $B = B_1 \cup B_2$ where $\varphi(a) < k$ when $a \in B_1$, $\varphi(a) = k$ when $a \in B_2$. We do need to consider B_1 any further.

For each $a \in B_2$, we choose x_a in the support of a, we let d_a be the distance of x_a to its complement in the support of a, and we let y_a be in the support of a, and such that $|x_a - y_a| = d_a$. We split $B_2 = B'_2 \cup B''_2$ where $d_a \geq 1$ on B'_2, $d_a < 1$ on B''_2.

If $a \in B'_2$, we put $a_1 = \varphi(x - x_a)a$, observe that a_1 is bounded independently of a, and has its support in x_a, while $a - a_1$ is also bounded independently of a and has $k - 1$ points in its support. This disposes of B'_2 by induction.

If $a \in B''_2$, we put $a_1 = \varphi((x - x_a)/d_a).a$. We define a'_1 from a_1 exactly as above, when k was equal to 2 and $a \in B''_2$. Then a'_1 ranges over an element of the boundedness generated by the sets $Y_{r.p}(K)$ while $a - a'_1$ ranges over a bounded subset of $F_r(U)$, with $\varphi(a - a'_1) \leq k-1$. This disposes of B''_2 by induction.

<u>Proposition 3</u> : Let $U \subseteq \mathbb{R}^n$. A mapping $f : U \to E$ belongs to $C_r(U, E)$ if, and only if it is possible to find, for $p \in \mathbb{N}^n$, $|p| \leq r$, continuous functions $f_p(x)$, $U \to E$ and $g_p(x,y)$, $U \times U \to E$ in such a way that $f_o(x) = f(x)$, $g_p(x,x) = o$ and

$$f_p(y) = \sum_{|q| \leq r-|p|} f_{p+q}(x) \frac{(y - x)^q}{q} + |y - x|^{r-|p|} g_p(x,y)$$

The system of functions f determines a linear mapping $\tilde{f}$ of $F_r(U)$ into E, which maps $i^{(p)}(x)$ on $f_p(x)$. We have $f = \tilde{f} \circ i$ if $f = f_o$. The mapping $\tilde{f}$ is in any case continuous on the sets $X_p(K)$. The relation between f and g, and the continuity of g ensure that $\tilde{f}$ is continuous on the sets $Y_{r.p}(K)$. It is therefore continuous on the elements of the compactology generated by the sets $X_p(K)$, $Y_{r.p}(K)$, i.e. $\tilde{f} \circ i = f$ is a mapping of class C_r of U into E.

<u>Proposition 3'</u> : The topology of $C_r(U, E)$ is equivalent to that determined by compact convergence of the functions f_p on U, and of the functions g_p on $U \times U$.

This is obvious, by now.

3. <u>The space $F_{r.s}(U, V)$</u>. A topology on the tensor product of two topological vector spaces $E \otimes F$ is admissible if the canonical bilinear mapping $E \times F \to E \otimes F$ is continuous. U and V will be two manifolds, each countable at infinity, $F_r(U)$ and $F_s(V)$ will be equipped with the direct limit topologies associated to their respective compactologies. In this section, we shall study the strongest admissible vector space topology on $F_r(U) \otimes F_s(V)$.

$\mathcal{E}_{r.s}(U, V)$ will be the injective tensor product $\mathcal{E}_r(U) \hat{\hat{\otimes}} \mathcal{E}_s(V)$. We consider this as a space of functions on $U \times V$. Modulo trivial identifications this is the space $C_r(U, \mathcal{E}_s(V))$, or again $C_s(V, \mathcal{E}_r(U))$. This is a locally convex Fréchet space. Its dual $\mathcal{E}^*_{r.s}(U, V)$ is a space of distributions. If $\varphi(a)$ denotes the number of elements of the support of an element of $\mathcal{E}^*_{r.s}$, then φ is weakly lower semi-continuous, $\varphi(a + b) \leq \varphi(a) + \varphi(b)$, and $\varphi(s\ a) = \varphi(a)$ for s scalar.

Definition 3 : $F_{r.s}(U, V)$ is the set of elements of $\mathcal{E}^*_{r.s}(U, V)$ whose support is finite. A subset B of $F_{r.s}(U, V)$ is bounded if B is equicontinuous and φ is bounded on B. We topologize each such set B with the weak topology $\sigma(\mathcal{E}^*_{r.s}(U, V), \mathcal{E}_{r.s}(U, V))$.

Clearly, $F_{r.s}(U, V)$ is a compactological space with a countable compactology.

The tensor product mapping maps $\mathcal{E}^*_r(U) \times \mathcal{E}^*_s(V)$ into $\mathcal{E}^*_{r.s}(U, V)$. If we put on $\mathcal{E}^*_r(U)$, $\mathcal{E}^*_s(V)$, and $\mathcal{E}^*_{r.s}(U, V)$ the topologies of uniform convergence on compact subsets of $\mathcal{E}_r(U)$, $\mathcal{E}_s(V)$ or $\mathcal{E}_{r.s}(U, V)$, it is known that $\mathcal{E}^*_{r.s}(U, V)$ is canonically identified with the completed, projective tensor product of $\mathcal{E}^*_r(U)$ and $\mathcal{E}^*_s(V)$.

Clearly, the tensor product mapping maps $F_r(U) \otimes F_s(V)$ into $F_{r.s}(U, V)$. The mapping is bijective here so that we can identify $F_{r.s}(U, V)$ with $F_r(U) \otimes F_s(V)$.

Proposition 4 : The topology of $F_{r.s}(U, V)$ is the strongest admissible vector space topology on $F_r(U) \otimes F_s(V)$.

It will be sufficient to show that the compactology of $F_{r.s}(U, V)$ is generated by the sets $B_1 \otimes B_2$ where B_1 is bounded in $F_r(U)$ and B_2 is bounded in $F_s(V)$. We may also assume that $U \subseteq \mathbb{R}^n$.

We let thus B be a bounded subset of $F_{r.s}(U, V)$. For each $b \in B$, we let $\varphi_1(b)$ be the number of elements of the projection of the support of b onto U, and $\varphi_2(b)$ be the number of elements of the projection of the support onto V. We assume that $\varphi_1(b) \leq k$ for all $b \in B$.

An induction on k, not unlike the one that was used in the proof of proposition 2 will give us the required result.

If $k = 0$, $B = \{0\}$ and there is nothing to prove.
If $k = 1$, if x_b is the point of the projection of its support in U, then

$$b = \sum i^{(p)}(x_b) \otimes c_{p.b}$$

where $c_{p.b} \in F_s(V)$ is defined by

$$\langle c_{p.b}, g\rangle = \langle b, (x - x_b)^p \otimes g(y)/p!\rangle$$

This shows such a set B is bounded in the boundedness generated by the sets $B_1 \otimes B_2$ where B_1 is bounded in $F_r(U)$ and B_2 is bounded in $F_s(V)$.

When $k > 1$, we may assume that the projections of the supports of the elements of B all have k elements. For each $b \in B$, we choose one of the points x_b of this projection.
We let d_b be the distance of x_b to its complement in the projection and choose y_b with $|x_b - y_b| = d_b$. We also choose a function ψ, of class C_∞, equal to unity on a neighbourhood of the origin in $\mathbb{R}^n$ and with compact support in the unit ball.

Then $B = B' \cup B''$ where $d_b \geq 1$ for $b \in B'$, $d_b < 1$ for $b \in B''$. For $b \in B'$, we let b_1 be the distribution with support in $\{x_b\} \times V$ which is equal to b on a neighbourhood of $\{x_b\} \times V$, then $b_1 = \varphi(x - x_b).b$. This shows that b_1 ranges over a bounded subset of $F_{r.s}(U \times V)$ when b ranges over B. The projection of the support of b_1 has a single point, that of the support of $b - b_1$ has $k - 1$ points. These remarks allow us to dispose of b_1 and of $b - b_1$ by induction, and show that B' is in the bounded structure generated by the sets $B_1 \otimes B_2$, B_1 bounded in $F_r(U)$, B_2 bounded in $F_s(V)$.

Let now $b \in B''$, and let b_1 be the distribution with support in $\{x_k\} \times V$ which is equal to b on a neighbourhood of $\{x_b\} \times V$. Then

$$b_1 = \psi\left(\frac{x - x_b}{d_b}\right) b$$

$$= \sum i^{(p)} (x_b) \otimes c_{p.b}$$

where the distribution $c_{p.b}$ is determined by the relation

$$\langle c_{p.b}, g\rangle = \langle b, \psi\left(\frac{x - x_b}{d_b}\right) \frac{(x - x_b)^p}{p!} \otimes g(y)\rangle$$

This shows that $d^{n-|p|} c_{pb}$ ranges over a bounded subset of $F_s(V)$ when b ranges over B'', so that

$$b'_1 = \sum (i^{(p)} (x_b) - T_{r.p}\, i^{(p)} (y_b, x_b)) \otimes c_{pb}$$

ranges over a set which is bounded in the boundedness generated by the sets $B_1 \otimes B_2$.

And $b - b'_1$ ranges over a set which is bounded in $F_{r.s}(U, V)$, the elements of this set have supports which project in U over finite sets having at most $k - 1$ elements, when $b \in B''$ of course.

This concludes the induction.

Proposition 5 : The space $C_r(U, C_s(V, E))$ can be identified with $\mathcal{L}(F_{r.s}(U, V), E)$.

$C_s(V, E)$ is the space of linear mappings $F_s(V) \to E$ whose restrictions to the bounded sets of $F_s(V)$ are continuous for the topology we consider on these bounded sets. $C_r(U, C_s(V, E))$ is the space of linear mappings $F_r(U) \to C_s(V, E)$ whose restrictions to the bounded sets of $F_r(U)$ are continuous.

$C_r(U, C_s(V, E))$ is therefore the space of bilinear mappings. $F_r(U) \times F_s(V) \to E$ whose restrictions to the sets $B_1 \times B_2$, B_1 bounded in $F_r(U)$ and B_2 in $F_s(V)$ are continuous. Applying proposition 4, we identify this space with the space of linear mappings $F_{r.s}(U, V) \to E$ whose restrictions to the bounded sets of $F_{r.s}(U, V)$ are continuous for the weak-star topology.

Corollary : $C_{r+s} (U \times V, E) \subseteq C_r (U, C_s(V, E))$

This follows from the fact that $F_{r.s}(U, V) \subseteq F_{r+s}(U \times V)$, the identity mapping being a morphism of compactological vector spaces.

4. Rapid approximation theorem. We shall start out by proving that the functions of class C_r, and finite ranks, are dense in $C_r(U, E)$, when E is a topological vector space. The proof will involve an explicit sequence of linear mappings of finite ranks of $F_r(U_1)$ in $F_r(U)$ where U_1 is open and relatively compact in U.

It will be sufficient to consider the case where $U \subseteq \mathbb{R}^n$. The general case (U a manifold) will be disposed of by means of a partition of unity subordinate to a covering of U by coordinate neighbourhoods.

c will be a function of class C_∞ on $\mathbb{R}^n$, with compact support in the open cube of side 2 and center at the origin, such that

$$\sum_{k \in \mathbb{Z}^n} c(x - k) = 1$$

For each $t > 0$, and each $x \in U_1$, we put

$$g_t(x) = \sum_{k \in \mathbb{Z}^n \cap t^{-1} U} c(t^{-1} x - k)\, T_r\, i(t.k, x)$$

Then g_t has finite rank, and belongs to $C_r(U_1, F_r(U))$. It extends therefore to a linear mapping $G_t : F_r(U_1) \to F_r(U)$.

Proposition 6 : G_t tends to the identity inclusion $F_r(U_1) \to F_r(U)$ uniformly on bounded (compact) subsets of $F_r(U_1)$ when $t \to 0$.

We observe that the support of $G_t\, ix$ has at most 2^n points, and that the support of $G_t\, a$ has at most $k.2^n$ points when the support of a has k elements. Let B be a bounded subset of $F_r(U_1)$, assume that the elements of B have supports with at most k elements. The elements of $G_t\, B$ have supports with at most $k.2^n$ elements.

We shall show that G_t tends to the identity immersion $F_r(U_1) \to \mathcal{E}^*_r(U)$, uniformly over B, when we put on $\mathcal{E}^*_r(U)$ the weak topology $\sigma(\mathcal{E}^*_r(U), \mathcal{E}_r(U))$. This will be sufficient for our purpose. If B is bounded subset of $F_r(U_1)$, and if the supports of its elements have at most k points, then the supports of the elements of $G_t B$ have at most $k.2^n$ points. Further, $\bigcup G_t B$ is weakly bounded in $\sigma(\mathcal{E}_r^*(U), \mathcal{E}_r(U))$, when we take the union over the range $0 < t < \varepsilon$, so this union is equicontinuous.

The closure of $\bigcup G_t B$ is a compact subset of $F_r(U)$. The topology induced on this set by $F_r(U)$ coincides therefore with that which $\mathcal{E}_r^*(U)$ induces. And proposition 6 follows.

To prove that G_t tends to the identity immersion $F_r(U_1) \to \mathcal{E}_r^*(U)$, it is sufficient to show that $\langle G_t a,u\rangle \to \langle a,u\rangle$ uniformly over bounded subsets of $F_r(U)$, this for all $u \in \mathcal{E}_r(U)$.

Keeping proposition 3', or proposition 2 in mind, we see that we must show that $u_t \to u$ in $\mathcal{E}_r(U_1)$ when $t \to 0$, if we put

$$u_t(x) = \sum_{k \in \mathbb{Z}^n \cap t^{-1} U} c(t^{-1} x - k)\, T_r u(t\, k,x)$$

but this is well known, and follows from the mean-value theorem.

<u>Corollary</u> : Elements of $C_r(U, E)$ which take their values in a finite dimensional subspace of E are dense in $C_r(U, E)$.

We must approximate $f \in C_r(U, E)$ by functions of class C_r and of finite ranks. This must be a C_r approximation on compact sets. We let K be compact in U, let u be a function of class C_∞ on U, equal to unity on a neighbourhood of K, and with compact support in U. We observe that fu approximates f ; it is therefore sufficient to approximate f on U_1, where U_1 is relatively compact in U.

$C_r(U, E) = \mathcal{L}(F_r(U), E)$. To prove the corollary, we must approximate the identity mapping $F_r(U_1) \to F_r(U)$ by mappings of finite ranks uniformly on compact subsets of $F_r(U_1)$; proposition 6 gives us the required system of approximants.

We now let $r_1 < r_2$, observe that $\mathcal{E}_{r_2}(U) \subseteq \mathcal{E}_{r_1}(U)$ is dense. Transposition gives us an injection $\mathcal{E}_{r_1}^* \to \mathcal{E}_{r_2}^*$ which maps

F_{r_1} into F_{r_2} and is a morphism of compactological spaces.

<u>Proposition 7</u> : We can associate to every bounded subset B of $F_{r_1}(U)$ a bounded sequence $u_1, \ldots, u_k, \ldots$ in $\mathcal{E}_{r_1}(U)$ and a bounded sequence $a_1, \ldots, a_k, \ldots$ in $F_{r_2}(U)$ in such a way that

$$\sum k^{-(r_2 - r_1)/n} \langle b, u_k \rangle a_k \to b$$

for all $b \in B$, and uniformly over B. The dimension of U is n.

We again assume that $U \subseteq \mathbb{R}^n$, B is bounded in $F_{r_1}(U)$, the supports of its elements are contained in a compact subset K of U. We let $N \in \mathbb{N}$, and write g_N instead of g_2-N in the proof of proposition 6.

$$g_N(x) - g_{N+1}(x) = \sum c(2^N x-k)\ c(2^{N+1} x-k')\ (T_r\, i(2^{-N} k,x) - T_r\, i(2^{-N-1} k',x))$$

We sum over the couples (k,k') such that $k \in 2^N U_1$, $k' \in 2^{N+1} U_1$. We may restrict the range of summation to the set of (k,k') such that $|k - 2k'| < 3M$, where M is the diameter of the cube of side 2, i.e. in the Euclidean norm $M = 2n^{1/2}$. This may be done because

$$c(2^N x - k) \,.\, c(2^{N+1} x - k') = 0$$

identically if $|k - 2k'| \geq 3M$.

We now decompose in an obvious way $g_N - g_{N+1}$ in a sum of terms, each of rank 1. The number of terms is $O(2^{nN})$.
Each of these terms is the product of $2^{-n(r_2 - r_1)}$ by some $u \otimes a$ where u ranges over a bounded subset of $F_{r_2}(U)$ and a over a bounded subset in $\mathcal{E}_{r_1}(U)$. The explicit computation uses the following identity over Taylor expansions.

$$T_{r_2}\, i(a,x) = \sum_{|p| \leq r_2} i^{(p)}(a) \frac{(x-a)^p}{p!}$$

$$= \sum_{|p_1 + p_2| \leq r_2} i^{(p)}(a) \frac{(x-b)^{p_1}}{p_1!} \frac{(b-a)^{p_2}}{p_2!}$$

$$= \sum_{|p| \leq r_2} T_{r_2 - |p|}\, i^{(p)}(a,b) \frac{(x-b)^p}{p!}$$

Since G_N tends to the identity when $N \to \infty$, the identity mapping can still be written

$$G_1 + \sum_1^\infty (G_{N+1} - G_N)$$

The above analysis shows that

$$G_{N+1} - G_N = \sum_{k_N+1}^{k_{N+1}} k^{-(r_2 - r_1)/n} \quad a_k \otimes u_k$$

where a_k is a bounded sequence in $F_{r_1}(U)$ and u_k is a bounded sequence in $F_{r_2}(U)$, while $k_N = 0\ (2^{-nN})$. To prove proposition 7, we must prove the following associativity result :

$$\sum_N \sum_{k_N+1}^{k_{N+1}} k^{-(r_2 - r_1)/n} \quad a_k \otimes b_k = \sum_k k^{-(r_2-r_1)/n} \quad a_k \otimes b_k$$

but this is clear, i.e.

$$\sum_{k_N+1}^{r} k^{-(r_2-r_1)/n} \quad a_k \otimes b_k \to 0$$

as $N \to \infty$, where we let r range over positive integers less than k_{N+1}.

5. Integration in locally pseudo-convex spaces. Propositions 6 and 7 are useful because they allow us to define the integrals of some vector-valued functions with range in locally pseudo-convex spaces.

Let E be a locally p-convex space. Assume that the bounded absolutely p-convex subsets of E are completant. Let μ be a measure with compact support in E. Choose r large enough so that $pr > n$.

Proposition 8 : A unique continuous linear mapping $C_{r-n}(U, E) \to E$ can be found, which maps the function $u(x).e$ where $u \in C_{r-n}(U)$ and $e \in E$ onto

$$\int u(x)\, d\mu(x) \,.\, e$$

Uniqueness follows from the density of $C_{r-n}(U) \otimes E$ in $C_{r-n}(U, E)$. The proof of the existence of a similar mapping of $C_r(U, E)$ into E is straightforward when we apply proposition 7. The function $f \in C_r(U, E)$ would be mapped onto

$$\sum k^{-r/n} \int \langle ix, u_k \rangle \, d\mu \,.\, \tilde{f}\, a_k$$

where $\tilde{f} : F_r(U) \to E$ is the continuous linear mapping, such that $f = \tilde{f} \circ i$. The series converges because E is locally p-convex and $\sum k^{-rp/n} < \infty$.

To prove the full strength of proposition 8, we must go into the proof of proposition 7. We have an increasing sequence of integers k_N where $k_N \to \infty$, $k_{N+1} - k_N = O(2^{mN})$. The functions u_k are bounded in the uniform norm, and if S_k is the support of u_k, then

$$\sum_{k_N+1}^{k_{N+1}} |\mu(S_k)| < K$$

where K is a constant depending only on m and μ.

The Hölder inequality now shows that

$$\sum_{k_N+1}^{k_{N+1}} |\mu(S_k)|^p = O(2^{mN(1-p)})$$

so that the "associated" series

$$\sum_N \sum_{k_N+1}^{k_{N+1}} k^{-r/n} \int \langle ix, u_k \rangle \, d\mu \; \tilde{f}\, a_k$$

converges. It is also clear, but not essential for our purposes, that the disassociated series

$$\sum_k k^{-r/N} \int \langle ix, u_k \rangle \, d\mu \, \tilde{f} \, a_k$$

also converges.

Definition 4 : The image of f by the continuous linear mapping whose existence is ensured by proposition 8 is called

$$\int f \cdot d\mu$$

Let now E be a locally pseudo-convex space. Assume that the closed absolutely convex subsets of E are completant. Let $\mathcal{B}$ be the small boundedness of E ($\mathcal{B}$ is generated by the closed absolutely convex hulls of rapidly decreasing sequences).

Proposition 9 : A bounded linear mapping $\tilde{f} : F(U) \to E$ has an unique bounded linear extension $\tilde{f}_1 : \mathcal{E}^*(U) \to E$.

We start out assuming that E is complete. Then E is an inverse limit of complete p-normed spaces. Proposition 8 allows us to define

$$\int f(x) \, d\mu(x)$$

whenever f is a function of class C_∞ on U, and μ a measure with compact support.

A distribution with compact support is a linear combination of derivatives of measures with compact supports.
If $f : U \to E$ is a function of class C_∞, if $\tilde{f} : F(U) \to E$ is such that $f = \tilde{f} \circ i$, we put

$$\tilde{f}_1 \left(\sum \frac{\partial^p \mu_p}{\partial x^p} \right) = \sum \int (-1)^{|p|} \frac{\partial^p f}{\partial x^p} \, d\mu_p$$

we must verify that

$$\sum (-1)^{|p|} \int \frac{\partial^p f}{\partial x^p} \, d\mu_p = 0$$

when $\sum \partial^p \mu_p / \partial x^p = 0$ but this is straightforward.

To end the proof we must investigate the case where E is not complete, but its closed, bounded, absolutely convex subsets are. The construction above maps $\mathcal{E}^*(U)$ into $E^\wedge$, the completion of E, and F(U) into E. The convex hulls of the bounded subsets of F(U) are bounded in $\mathcal{E}^*(U)$ and are therefore mapped onto bounded subsets of E.

It is clear that the bounded subsets of $\mathcal{E}^*(U)$ are contained in the completant hulls of the bounded subsets of F(U) (or of the convex hulls of these bounded subsets). The linear mapping $\mathcal{E}^*(U) \to E^\wedge$ maps $\mathcal{E}^*(U)$ into the union of the completant hulls of the convex bounded subsets of E, but these hulls are contained in E. So $\tilde{f}_1$ is a bounded linear mapping of $\mathcal{E}^*(U)$ into E.

<u>Corollary</u> : The extension $\tilde{f}_1 : \mathcal{E}^*(U) \to E$ is a bounded linear mapping when we consider on $\mathcal{E}^*(U)$ the equicontinuous boundedness, and on E the small boundedness.

This is clear. The equicontinuous subsets of $\mathcal{E}^*(U)$ are contained in the completant absolutely convex hulls of rapidly decreasing sequences because this is a nuclear b-space (cf. [6]). The images of these equicontinuous sets are small bounded sets.

Let $(E,\mathcal{B})$ be a b-space, and U a manifold. A mapping $f : U \to E$ is differentiable (of class C_∞) if it is possible to find, for each compact $K \subseteq U$ and each $r \in \mathbb{N}$ a completant bounded set B in such a way that the restriction of f to some neighbourhood U_1 of K is a mapping of class C_r of K into E_B (cf. [86]). The following is clear :

<u>Corollary</u> : Let $(E,\mathcal{T})$ be a locally pseudo-convex space whose closed absolutely convex bounded sets are completant ; let $\mathcal{B}$ be the small boundedness of $(E,\mathcal{T})$. A mapping $f : U \to E$ is of class C_∞ bornologically if it is of class C_∞ topologically.

<u>Notes and remarks</u> : The origin of the research in this chapter is a joint research by P. Turpin and the author, about integration in locally pseudo-convex spaces, holomorphic vector valued functions and locally pseudo-convex algebras (Ref. [74], [75], [76], [87]). An earlier draft of these results is contained in a set of lecture notes by the author [84].

The present lecture notes will not contain the results about holomorphic functions, nor do we give here the examples which show that the condition $pr \geq n$ is essential if we wish to prove proposition 8. The earlier set of lecture notes contained at least one error. The corollary of proposition 5, here, was considered obvious.

The results related to topological algebras will be given further (see for instance chapter V, paragraph 4).

At the time Turpin and the author first met, Turpin had managed to integrate Lipschitz functions taking their values in a complete p-normed space, with $p > 1/2$. This was possible, because the Lipschitz condition gave him a satisfactory control on the error term in the finite sums which tend to the integral. Discussion showed that smaller values of p could be handled if one tried to integrate smoother functions, and it was completely reasonable to adopt Whitney's definition, and define a function of class C_r as is done in proposition 3. It was only later that the author noticed the equivalence of Whitney's definition with the definition we use here.

Other authors had integrated functions taking their values in p-normed, locally p-convex, and locally pseudo-convex spaces. B. Gramsch [31], and independently D. Przeworska-Rolewicz and S. Rolewicz [64] studied the projective tensor product for the exponent p of $C(X)$, X compact, and a p-normed space, E. They showed that the natural mapping $C(X) \hat{\otimes}_p E \to C(X, E)$ is injective. This supplies them with a broad class of integrable functions. The functions of class C_r that we can integrate belong to $C(X) \hat{\otimes}_p E$ (proposition 8) or to $L_1(X) \hat{\otimes}_p E$ (proposition 8).

The space $L_1 \hat{\otimes}_p E$ was studied by D. Vogt [78], who showed that its elements can be identified with special almost everywhere defined functions.

Another approach to integration of vector valued functions was initiated by D.O. Etter Jr. [19]. Etter's definition was dissected by the author ([85], paragraph 7). B. Gramsch on the other hand showed that Etter had considered the injective tensor product of a special Banach space and a topological vector space.

He studied the injective tensor product of a locally convex space and a topological vector space, and the relation between this injective tensor product and the projective tensor product (Ref. [32], [34]).

As far as tensor products of topological vector spaces are concerned, it must be said that many problems remain open. Let E and F be Hausdorff topological vector spaces. Does there exist a Hausdorff admissible topology on $E \otimes F$? It is the case if E or F can be embedded continuously in a product of spaces such as M(I), the almost everywhere defined functions with convergence in measure.

Paragraph 3 contains a minor result related to topological tensor products of topological vector spaces.

CHAPTER V

The Gelfand-Mazùr theorem

1. Banach algebras : We shall usually be considering commutative algebras, with unit and complex coefficients. The fact will be mentioned explicitly when this is not the case. The unit will be identified with the complex number 1, the product of the unit with the complex number s will be identified with s.
The complex field becomes in this way a subalgebra of the given algebra.

There are standard means of adapting results about algebras with unit, and applying them to algebras without unit. We shall not go back over these. Let us recall that they involve adjoining a unit and removing it, regular (maximal) ideals, quasi-inverses, etc.

A Banach algebra is an algebra $\mathfrak{a}$, with a Banach space norm, which satisfies the further condition

$$||a.b|| \leq ||a|| \cdot ||b||$$

$$||1|| = 1$$

It is well known that these two conditions are essentially equivalent to the continuity of multiplication. If multiplication is continuous in a normed algebra, $\mathfrak{a}$, then $\mathfrak{a}$ possesses an equivalent norm satisfying the two conditions above, or only the first one if $\mathfrak{a}$ does not have a unit.

Proposition 1 : The set of invertible elements of a Banach algebra is open. The mapping $a \to a^{-1}$ is continuous on its domain.

This is well known. If $||x|| < 1$, we have

$$(1 - x)^{-1} = 1 + x + x^2 + \dots \qquad (1)$$

This series converges uniformly for $||x|| \leq 1 - \varepsilon$. This already shows that a^{-1} is defined and continuous for a near to the unit.

If a has an inverse, and if $||h|| < ||a^{-1}||^{-1}$,

$$(a - h)^{-1} = (1 - a^{-1} h)^{-1} a^{-1} \qquad (2)$$

<u>Proposition 2</u> : The mapping $a \to a^{-1}$ is even analytic in the sense of E.R. Lorch [51].

This analyticity means that it is locally the sum of a power series in a single $\mathfrak{a}$-variable. Proposition 1 holds whether $\mathfrak{a}$ be commutative or not. Proposition 2 holds when $\mathfrak{a}$ is commutative. Formula (1) gives a power series expansion for a^{-1} around the unit, formula (2) allows us to obtain a power series expansion for $(a - h)^{-1}$ around $h = 0$, this when a and h commute.

We observe that $a \to a^{-1}$ is in any case an analytic mapping of the open set of invertible elements into itself, when we are interested in usual analytic mappings of complex manifolds (with complex coefficients) rather than $\mathfrak{a}$-analytic mappings, this whether $\mathfrak{a}$ is commutative or not.

<u>Definition 1</u> : Let $\mathfrak{a}$ be a Banach algebra, let $a \in \mathfrak{a}$. The spectrum of a, sp a, is the set of complex numbers s such that $a - s$ does not have an inverse.

<u>Proposition 3</u> : The spectrum of a Banach algebra element is a compact, non empty subset of the complex plane. The resolvent function $h \to (a - h)^{-1}$ is analytic on the complement of the spectrum and vanishes at infinity.

The spectrum is closed because $s \notin \text{sp } a$ iff $(a - s)$ has an inverse and the set of invertible elements is open. It is bounded, because $(a - s) = -s(1 - s^{-1}a)$ has an inverse if $|s| > ||a||$. So the spectrum is compact.

Analyticity of the resolvent $(a - s)^{-1}$ for $s \neq \infty$ follows by restriction from the analyticity of a^{-1} on its domain. It follows at infinity from the relation

$$(a - s)^{-1} = -s^{-1}(1 - s^{-1}a)^{-1}$$

Liouville's theorem shows that the spectrum is not empty because the resolvant is not constant.

<u>Proposition 4</u> : (Gelfand-Mazùr) The only Banach algebra that is a field is the complex field.

Assume $\mathfrak{a}$ is a Banach algebra, and a field. Let $a \in \mathfrak{a}$, and $s \in \operatorname{sp} a$. Then $a - s$ does not have an inverse, and $a = s \in \mathbb{C}$, i.e. $\mathfrak{a} \subseteq \mathbb{C}$. But $\mathbb{C} \subseteq \mathfrak{a}$.

An ideal α of a Banach algebra is a true ideal if $\alpha \neq \mathfrak{a}$. Maximal elements of the set of true ideals are interesting. These are called maximal ideals.

<u>Proposition 5</u> : Every true ideal is contained in a maximal ideal. Maximal ideals are closed.

The first statement follows from the fact that α is a true ideal iff $1 \notin \alpha$, so Zorn's lemma can be applied. The second statement follows from the observation that α does not meet the set of invertible elements if α is true, but the set of invertible elements is a neighbourhood of the unit, so the closure of a true ideal is true. The closure of a maximal ideal is a larger, true ideal.

<u>Proposition 6</u> : The quotient of a Banach algebra with respect to a maximal ideal is the complex field. If m is a maximal ideal, and a an element of the algebra, then a complex number $\hat{a}(m)$ can be found such that $a - \hat{a}(m) \in m$.

The first statement is a corollary of the Gelfand-Mazùr theorem. The quotient norm on $\mathfrak{a}/m$ is a Banach algebra norm. The quotient of a ring with unit with respect to a maximal ideal is a field. So $\mathfrak{a}/m = \mathbb{C}$, up to a natural isomorphism.

The second statement is a rewording of the first. Write $a - \hat{a}(m) \in m$ rather than $a - \hat{a}(m) \equiv 0 \pmod{m}$.

<u>Definition 2</u> : Let $\mathfrak{a}$ be a Banach algebra, $\mathfrak{m}$ the set of its maximal ideals. $\mathfrak{T}$ will be the weakest topology on $\mathfrak{m}$ such that the functions $\hat{a}$ are continuous functions on $\mathfrak{m}$ for all $a \in \mathfrak{a}$.

Proposition 7 : $\mathcal{T}$ is a compact topology. The mapping $a \to \hat{a}$ is a continuous homomorphism of $\mathfrak{a}$ into $C(\mathfrak{M})$.

We associate to every $m \in \mathfrak{M}$ the mapping $a \to \hat{a}(m)$, $\mathfrak{a} \to \mathbb{C}$. This is a multiplicative linear form whose kernel is m. Conversely the kernel of a multiplicative linear form is a maximal ideal. Multiplicative linear forms are continuous, because their kernels are closed.
Better, if χ is a multiplicative linear form, and $a \in \mathfrak{a}$, then $a - \chi(a)$ does not have an inverse, so $|\chi(a)| \leq ||a||$.

We have defined in this way a bijection between $\mathfrak{M}$ and a weakly closed subset of the unit ball of $\mathfrak{a}^*$, the topological dual of $\mathfrak{a}$. The image of $\mathfrak{M}$ is $\sigma(\mathfrak{a}^*, \mathfrak{a})$-compact. And $\mathcal{T}$ is the topology on $\mathfrak{M}$ for which this bijection is a homeomorphism with its image, the image being topologized by $\sigma(\mathfrak{a}^*, \mathfrak{a})$.

The second statement is a triviality. Continuity of the mapping follows from the inequality $||\hat{a}||_\infty \leq ||a||$.

Definition 3 : Let $\mathfrak{a}$ be a commutative Banach algebra with unit. Let $(a_1, \ldots, a_n)$ be elements of $\mathfrak{a}$. The joint spectrum of $(a_1, \ldots, a_n)$ is the set of $(s_1, \ldots, s_n) \in \mathbb{C}^n$ such that $(a_1 - s_1, \ldots, a_n - s_n)$ generates a true ideal of $\mathfrak{a}$. We call this set $sp(a_1, \ldots, a_n)$.

Proposition 8 : $sp(a_1, \ldots, a_n)$ is a non empty compact subset of $\mathbb{C}^n$. It is the set

$$sp(a_1, \ldots, a_n) = \{(\hat{a}_1(m), \ldots, \hat{a}_n(m)) \mid m \in \mathfrak{M}\}$$

What has to be proved is that $(a_1 - s_1, \ldots, a_n - s_n)$ generate a true ideal iff some $m \in \mathfrak{M}$ can be found such that $s_i = \hat{a}_i(m)$ for all m. But if these elements generate a true ideal, then this ideal is contained in some maximal ideal m, and $s_i = \hat{a}_i(m)$. And if $s_i = \hat{a}_i(m)$ for all i, $a_i - s_i = a_i - \hat{a}_i(m) \in m$; the generated ideal is contained in m.

The joint spectrum is not empty because $\mathfrak{M}$ is not empty. It is compact because it is the image of a compact set by a continuous mapping.

Definition 4 : The spectral radius of $a \in \mathfrak{a}$ is defined by

$$\rho(a) = \max \{|s| \mid s \in \operatorname{sp} a\}$$

In other words, $\rho(a)$ is the radius of the smallest circle centered at the origin and containing sp a.

Since $\operatorname{sp} a = \hat{a}(\mathfrak{M})$, we see that $\rho(a) = ||\hat{a}||_\infty$.

Proposition 9 : The spectral radius is a submultiplicative semi-norm on $\mathfrak{a}$. Its kernel is the intersection of the maximal ideals of $\mathfrak{a}$ (the radical of $\mathfrak{a}$).

This is obvious when we remember that $a \to \hat{a}$ is a homomorphism of $\mathfrak{a}$ into $C(\mathfrak{M})$, and that $\rho(a) = ||\hat{a}||_\infty$. Also, $\rho(a) = 0$ iff $\hat{a} = 0$ identically on $\mathfrak{M}$, but when this is the case a is in the intersection of the maximal ideals, and conversely.

Proposition 10 : The sequence $||a^n||^{1/n}$ tends to a limit, equal to the spectral radius, when $n \to \infty$.

Let n be given, let m be large, write $m = nk + r$ with $0 \leq r < n$. Since $||a^{nk}|| \leq ||a^n||^k$, we see that $||a^{nk}||^{1/nk} \leq ||a^n||^{1/n}$ and that

$$||a^m||^{1/m} \leq (||a^{nk}||^{1/nk})^{nk/(nk + r)} \cdot ||a^r||^{1/m}$$

$$\leq (||a^n||^{1/n})^{nk/(nk + r)} \cdot ||a^r||^{1/m}$$

This shows that we can associate m_o to ε in such a way that

$||a^m||^{1/m} \leq ||a^n||^{1/n} + \varepsilon$ whenever $m \geq m_o$. Hence

$$||a^n||^{1/n} \geq \limsup_{m \to \infty} ||a^m||^{1/m}$$

for all n, and the limit exists.
The fact that the limit is the spectral radius follows from the fact that the series

$$(a - z)^{-1} = - \sum_0^\infty z^{-k-1} a^k$$

converges on the complement of the smallest circle centered at the origin which contains all the singularities of the function, i.e. it converges if $|z| > \rho(a)$ and diverges if $|z| < \rho(a)$, but it converges if $|z| > \lim ||a^n||^{1/n}$ and diverges if $|z| < \lim ||a^n||^{1/n}$.

2. Regular elements in b-algebras. We shall now assume that $\mathfrak{a}$ is a commutative b-algebra with unit. The results proved in this section would be relatively easy to generalize to the case where $\mathfrak{a}$ does not have a unit. But nothing very interesting could be said when $\mathfrak{a}$ is not commutative.

Definition 5 : An element a of $\mathfrak{a}$ is regular if the resolvent function $(a - s)^{-1}$ is defined and bounded on a neighbourhood of infinity.

Definition 5' : The regular resolvent set of the regular element a is the set of complex numbers such that $(a - s)^{-1}$ exists and is a regular element. The regular spectrum of a is the complement of the regular resolvent set. The regular spectrum of a will be called $sp_r\ a$.

Proposition 11 : The element a is regular if and only if some real number M can be found such that a^n/M^n is a bounded sequence.
The regular resolvent set of the regular element a is the largest open set on which the resolvent is locally bounded.
The resolvent is analytic on this set, and has at infinity the expansion

$$(a - s)^{-1} = - \sum_0^\infty s^{-n-1}\ a^n$$

The regular spectrum of a regular element is compact and not empty.
If we put $\rho_r(a) = \max \{|s| \mid s \in sp_r\ a\}$, we have

$$\rho_r(a) = \inf \{M \mid M > 0,\ a^n/M^n \text{ is bounded}\}$$

We must first see that the regular resolvent set is the largest open set on which the resolvent is locally bounded. This follows from the two identities

$$(a - s_o - h)^{-1} = (a - s_o)^{-1} - (a - s_o)^{-2} \left[(a - s_o)^{-1} - h^{-1}\right]^{-1}$$

$$\left[(a - s_o)^{-1} - h^{-1}\right]^{-1} = (a - s_o)\ h\ (a - s_o - h)^{-1}$$

The first shows that $(a - s)^{-1}$ is bounded on a neighbourhood of s_o if $(a - s_o)^{-1}$ is regular. The second one shows that $(a - s_o)^{-1}$ is regular if $(a - s)^{-1}$ is defined and bounded on a neighbourhood of s_o.

We must show that the resolvent is analytic on the regular resolvent set. If U is a complex domain and E a b-space, we say that a mapping $f : U \to E$ is analytic if for each $z \in U$ we can find a neighbourhood V of z and a completant bounded set B of E in such a way that f is a holomorphic mapping of V into E_B. The analyticity of the resolvent follows from the resolvent identity.

$$(a - s - h)^{-1} - (a - s)^{-1} = h\,(a - s)^{-1}\,(a - s - h)^{-1}$$

This shows that the resolvent maps V continuously into $\mathfrak{a}_B$ if B is bounded, completant, and if B is chosen in such a way that $(a - s)^{-1}\,(a - s - h)^{-1} \in B$ when $s \in V$, $s + h \in V$.

If we now divide by h, and assume that $(a - s_1)^{-1}\,(a - s_2)^{-1}\,(a - s_3)^{-1} \in B$ when $s_1 \in V$, $s_2 \in V$, $s_3 \in V$, we see that $(a - s)^{-1}$ is a complex-differentiable function taking its values in $\mathfrak{a}_B$, the derivative being equal to $(a - s)^{-2}$.

The resolvent is holomorphic on a neighbourhood of infinity and bounded on that neighbourhood. It possesses a power series expansion $\sum c_n z^{-n}$ near to infinity and $(a - s) \sum c_n z^{-n} = 1$ identically.

This yields an inductive system of equations for c_n, which is easy to solve

$$(a - s)^{-1} = - \sum a^n \; s^{-n-1}$$

Proof of proposition 11 is now straightforward. The fact that a is regular if a^n/M^n is bounded is clear since $\sum z^{-n-1} a^n$ converges uniformly for $|z| > M + \varepsilon$. And conversely, if $(a - z)^{-1}$ is defined and bounded for $|z| > M$, it is classical that a^n/M^n is a bounded sequence.

Proposition 12 : Let $\mathfrak{a}$ be a commutative b-algebra with unit. The set $\mathfrak{a}_r$ of regular elements of $\mathfrak{a}$ is a subalgebra. The spectral radius is a multiplicative semi-norm on $\mathfrak{a}_r$. If $\mathfrak{m}$ is the set of maximal ideals of $\mathfrak{a}_r$, then to every $a \in \mathfrak{a}_r$ and $m \in \mathfrak{m}$, we can associate a complex number $\hat{a}(m)$ such that $a - \hat{a}(m) \in m$. The weakest topology on $\mathfrak{M}$ for which the functions $\hat{a}$ all are continuous is a compact topology. The regular spectrum of $a \in \mathfrak{a}_r$ is given by

$$sp_r\ a = \hat{a}(\mathfrak{M})$$

It is trivial that sa is regular, and that $\rho(sa) = |s|\ \rho(a)$ if a is regular and s is scalar. If a and b are regular, if $M > \rho(a)$ if $N > \rho(b)$, then $(ab)^n/(MN)^n$ is a bounded sequence and $M.N \geq \rho(a.b)$ so $a.b$ is regular and $\rho(a.b) \leq \rho(a)\ \rho(b)$.

To show that $a + b$ is regular and $\rho(a + b) \leq \rho(a) + \rho(b)$, we must prove that $(a + b)^n/(M + N)^n$ is a bounded sequence, but

$$\frac{(a + b)^n}{(M + N)^n} = \sum_k \binom{n}{k} \frac{M^k\ N^{n-k}}{(M + N)^n}\ \frac{a^k}{M^k}\ \frac{b^{n-k}}{N^{n-k}}$$

These are convex combinations of elements of a bounded set.

The spectral radius ρ is a multiplicative semi-norm on $\mathfrak{a}_r$. It determines a (non Hausdorff) topology on this algebra. Multiplication is continuous in that topology. The ρ-closure of a true ideal is a true ideal, this follows from the fact that $1 - a$ has an inverse in $\mathfrak{a}_r$ if $\rho(a) < 1$, so that the unit has a ρ-neighbourhood made up of invertible elements in $\mathfrak{a}_r$. Maximal ideals m of $\mathfrak{a}_r$ are ρ-closed and ρ induces a Banach algebra norm on the field $\mathfrak{a}_r/m$. We may apply the Gelfand-Mazùr theorem.

The end of the proof is straightforward, follows from the same types of development as those found in the preceding paragraph.

A subset X of an algebra $\mathfrak{a}$ is idempotent if $X^2 \subseteq X$. The convex hull of an idempotent set is idempotent.

<u>Definition 6</u> : A subset B of a b-algebra $\mathfrak{a}$ is bounded for the Allan boundedness of $\mathfrak{a}_r$ if $B \subseteq MB_1$ with $M \in \mathbb{R}_+$ and B_1 bounded idempotent.

<u>Proposition 13</u> : $\mathfrak{a}_r$ is the union of the sets MB with $M \in \mathbb{R}_+$, and B bounded idempotent. The Allan boundedness determines a b-algebra structure on $\mathfrak{a}_r$.

The first statement in proposition 13 just ensures that the Allan boundedness is a boundedness on $\mathfrak{a}_r$. It is obvious ; if $x \in MB$, with $M \in \mathbb{R}_+$ and B bounded idempotent, then $\{x^n/M^n \mid n \in \mathbb{N}\}$ is bounded and B is regular ; if x is regular, then $B = \{x^n/M^n \mid n \in \mathbb{N}\}$ is bounded for M large, it is idempotent and $x \in MB$.

To show that the Allan boundedness is a b-algebra boundedness, we must show that a bounded idempotent set is contained in a bounded completant idempotent set. We know that the absolutely convex hull of a bounded idempotent set is bounded and idempotent. We must show that the completant hull of a bounded absolutely convex idempotent set is idempotent.

Let B be bounded, absolutely convex, and idempotent. Let B_1 be its completant hull. Let x and y be elements of B_1, and x_n, y_n be sequences of elements of B with $x_m - x_n \in \varepsilon_n B$, $y_m - y_n \in \varepsilon_n B$ when $m \geq n$, with $x_n \to x$, $y_n \to y$. Then $x_n y_n \to x\,y$, and $x_n\,y_n - x_m\,y_m \in 2\varepsilon_n\,B$ when $m \geq n$. (The convergence of the sequences $x_n \to x$, $y_n \to y$, $x_n\,y_n \to xy$ are for the convergence structure associated to the boundedness of $\mathfrak{a}$).

<u>Proposition 14</u> : The union of two bounded idempotent sets is contained in a bounded idempotent set.

If B is idempotent, then $B \cup \{1\}$ is idempotent and contains the unit. If B_1 and B_2 are idempotent, each containing the unit, then $B_1.B_2$ is idempotent and contains $B_1 \cup B_2$.

For each B, completant and idempotent, $\mathfrak{a}_B$ is a Banach algebra. This Banach algebra has a unit if $1 \in B$. We can define $\mathfrak{m}_B$, the space of maximal ideals of $\mathfrak{a}_B$.
If $B' \supseteq B, \mathfrak{a}_{B'} \supseteq \mathfrak{a}_B$. The restriction mapping is continuous $\mathfrak{m}_{B'} \to \mathfrak{m}_B$. It is therefore possible to speak of the projective limit of the compact spaces $\mathfrak{m}_B$, for these structural mappings, when B ranges over the bounded, completant, idempotent subsets of $\mathfrak{a}$ which contain the unit. This is a projective limit of compact spaces. Such a projective limit behaves well.

<u>Proposition 15</u> : The projective limit of the spaces $\mathfrak{m}_B$ is the space $\mathfrak{m}$ of maximal ideals of the algebra $\mathfrak{a}_r$.

A maximal ideal m of $\mathfrak{a}_r$ induces a maximal ideal m_B of $\mathfrak{a}_B$ for each B bounded, completant, idempotent, $1 \in B$. This system of maximal ideals is an element of the projective limit. And if $\{m_B\}$ is an element of the projective limit, $m = \bigcup m_B$ is an ideal of $\mathfrak{a}_r = \bigcup \mathfrak{a}_B$, and $\mathfrak{a}/m$ is isomorphic to $\mathbb{C}$, so that m is maximal in $\mathfrak{a}_r$.

Note : We shall define later a weaker boundedness on $\mathfrak{a}_r$, the equiregular boundedness. We shall show that the Allan boundedness is the set of sets which have an equiregular convex hull.

3. Continuous inverse algebras. Definition 7 : $\mathfrak{a}$ is a continuous inverse algebra if $\mathfrak{a}$ is a topological algebra, with jointly continuous multiplication, and such that a^{-1} is defined and continuous on a neighbourhood of the unit.

We shall see later (chapter VIII, paragraph 3) that a locally convex commutative continuous inverse algebra is "locally multiplicatively convex", i.e. that its topology can be determinded by a system of multiplicative semi-norms.

Proposition 16 : Let $\mathfrak{a}$ ba an algebra, $\mathcal{T}$ a vector space topology on $\mathfrak{a}$. Assume that the set of invertible elements of $\mathfrak{a}$ is a neighbourhood of the unit, and that $a^{-1} \to 1$ when $a \to 1$. The set of invertible elements of $\mathfrak{a}$ is then open, a^{-1} is a continuous function of a on its domain, and the Jordan multiplication of $\mathfrak{a}$, $(a,b) \to a.b + b.a$ is continuous.

The proof of this proposition follows from a number of small identities.

$$x^2 = [x^{-1} - (x + 1)^{-1}]^{-1} - x$$

so $x^2 \to 1$ as $x \to 1$ and by polarization, $x.y + y.x$ is continuous.

$$2\,x\,y\,x = [x(x\,y + y\,x) + (x\,y + y\,x)x] - [x^2\,y + y\,x^2]$$

is again a continuous function of (x,y).

Let now x be invertible in $\mathfrak{a}$, let $h \in \mathfrak{a}$ be small, so that $u = h\,x + x\,h + h^2$ is small. Then

$$x^2 + u = x(1 + x^{-1}\,u\,x^{-1})x$$

has an inverse, and this inverse is near to x^{-2}. But $x^2 + u = (x + h)^2$ and

$$2(x + h)^{-1} = (x + h)\,(x + h)^{-2} + (x + h)^{-2}\,(x + h)$$

We have obtained the result, that y^{-1} is defined when y is near to x, and that $y^{-1} \to x^{-1}$ when $y \to x$.

Proposition 16 allows us to give equivalent, but apparently weaker definitions of continuous inverse algebras, especially when $\mathfrak{a}$ is commutative, so that multiplication and Jordan multiplication do not differ essentially.

Proposition 17 : Let $\mathfrak{a}$ be a continuous inverse algebra, let $a \in \mathfrak{a}$. The set of $s \in \mathbb{C}$ such that $(a - s)^{-1}$ does not exist is compact. The resolvent function $(a - s)^{-1}$ is of class C_∞ on the complement of this set, and at infinity. It is a solution of the Cauchy-Riemann equations.

We must observe that the "spectrum of a" may be empty. This is due to the fact that we do not make any convexity hypothesis on the topology of $\mathfrak{a}$. Liouville's theorem cannot be applied. Chapter IX will be devoted to the study of topological extensions of $\mathbb{C}$, a true topological extension must contain many elements with "empty spectrum", as the proof of the Gelfand-Mazùr theorem shows.

We want to show that $(a - s)^{-1}$ is of class C_r for every integer r. We let $f_o(s) = (a - s)^{-1}$, and $f_k(s) = k!\,(a - s)^{-k-1}$. The geometric expansion formula shows that

$$f_o(s + h) = \sum_o^r f_k(s)\,\frac{h^k}{k!} + h^{r+1}\,(a - s)^{-r}\,(a - s - h)^{-1}$$

$$= \sum_o^r f_k(s)\,\frac{h^k}{k!} + o(h^r)$$

This shows that our candidate Taylor expansion of $f_o(s) = (a - s)^{-1}$ is near enough to $(a - s - h)^{-1}$ to qualify. Before being satisfied, however, we must investigate $f_p(s) = p!\,(a - s)^{-p-1}$ and show that

$$f_p(s + h) = \sum_{|k| \leqslant r-|p|} f_{k+p}(s)\,\frac{h^k}{k!} + o(|h|^{r-|p|}).$$

In other words, that

$$(a - s - h)^{-(p+1)} = \sum \binom{k+p}{k} (a - s)^{-k-p-1} h^k + o(h^{r-|p|})$$

and this follows from raising the geometric expansions of $(a - s - h)^{-1}$ to the power $p+1$, grouping together the terms involving h^k for $0 \leq k \leq r - |p|$ and lumping together the higher order terms. This shows that the resolvent is of class C_r for every r, i.e. it is of class C_∞ (applying proposition 3, chapter IV).

A function is a solution of the Cauchy-Riemann equations if the linear term of its Taylor expansion is complex linear. This is the case here.

The compactness of the set of s such that $(a - s)^{-1}$ does not exist is nearly trivial : the proof that the spectrum is compact, in proposition 3, only uses the fact that the set of invertible elements is open.

We must still show that $(a - s)^{-1}$ is of class C_∞ at infinity i.e. that $(a - s^{-1})^{-1}$ is of class C_∞ at the origin. This may be done by observing that

$$(a - s^{-1})^{-1} = - s\ (1 - s\ a)^{-1}$$

and showing that $(1 - s\ a)^{-1}$ is of class C_∞ at the origin. This is not essentially different from the proof above, so that we shall not repeat the developments here.

<u>Proposition 18</u> : Let $(\mathfrak{a}, \mathfrak{T})$ be a continuous inverse locally pseudo-convex algebra, and let $\mathfrak{B}$ be its small boundedness. Assume that $\mathfrak{B}$ is a completant boundedness. Every element of $\mathfrak{a}$ is a regular element of the b-algebra $(\mathfrak{a}, \mathfrak{B})$.

Since $(a - s)^{-1}$ is a function of class C_∞ on its domain, as $(\mathfrak{a}, \mathfrak{T})$-valued function, we see that it is also an $(\mathfrak{a}, \mathfrak{B})$-valued function of class C_∞ on its domain (chapter IV, proposition 9), so that $(a - s)^{-1}$ is locally bounded at infinity.
This shows that a is a regular element of $(\mathfrak{a}, \mathfrak{B})$.

Proposition 18 allows us to apply the results of the preceding paragraph to continuous inverse algebras, when they are locally pseudo-convex. The spectra of their elements are not empty. Their spectral radius can be defined,

$$\rho(a) = \max \{|s| \mid s \in \operatorname{sp} a\}$$

$$= \inf \{M > 0 \mid a^n/M^n \text{ is bounded}\}$$

<u>Proposition 19</u> : The spectral radius is a continuous semi-norm.

This is true, since $\rho(a) < 1$ when a belongs to a balanced neighbourhood of the origin such that $(1 + x)^{-1}$ exists for all x in that neighbourhood.

<u>Note</u> : In the proof of proposition 17, it is sufficient to assume that $(a - s)^{-1}$ is defined and bounded on an open subset of $\mathbb{C}$. The resolvent function will then be of class C_∞, and a solution of the Cauchy-Riemann equations on that set. Of course, when we say that $(a - s)^{-1}$ is bounded, we mean that it is bounded for some boundedness on $\mathfrak{a}$, which is multiplicative. A mapping $f : U \to E$ of a manifold into a vector space on which a boundedness is defined is of class C_∞ if $f = f_1 \circ i$ where $f_1 : F(U) \to E$ is a bounded linear mapping, and $i : U \to F(U)$ is the natural embedding.

<u>Notes and remarks</u> : The results of paragraph 1 were proved by I.M. Gelfand in 1941 [29]. The Gelfand-Mazùr theorem had been announced previously by S. Mazùr [52]. Mazùr's proof was never published, it involved consideration of harmonic functions rather than analytic functions. The material is classical by now. It is included as a backdrop, before developping less classical aspects of the theory.

Paragraph 2 is an adaptation of results of the author [80] and of G. Allan [4]. Allan does not really assume that the topological algebras he studies are complete, sequentially complete, or anything of the sort, but he does assume that the bounded closed absolutely convex idempotent sets are completant, so that the Allan boundedness is completant.

When the regular elements were introduced, it was hoped that they could be thought of as bounded functions, or bounded operators.

The non-regular elements (or some of these) would be interpreted as unbounded functions, unbounded, densely defined operators, and could be studied starting out from the regular elements.

Up to now, at least, the author does not believe that this hope has been substantiated. The algebra $\mathfrak{a}_r$ however is a subalgebra of $\mathfrak{a}$, whose elements are "nice" in a well defined way. It deserves consideration.

R. Arens [6] has proved that a continuous inverse locally convex algebra cannot be a field, unless it is $\mathbb{C}$ itself. As a matter of fact Arens proved a stronger result, which we shall investigate in chapter IX. Once this theorem has been obtained, extension of the results in paragraph 1 to continuous inverse locally convex algebras is straightforward.

W. Żelazko investigated complete locally bounded algebras [95], [96], [97], [98]. He proved that many results which are standard in Banach algebra theory could be generalized to such algebras.

Of course, these algebras are p-normable for some p, because of the Aoki-Rolewicz theorem (Chapter I, proposition 3, corollary 1), and the standard proof, valid in Banach algebra theory, shows that they are continuous inverse algebras.

Żelazko's proofs were difficult, because he did not integrate. And he did not integrate because the vector spaces in which the functions he considered take their values are not locally convex. S. Mazur and W. Orlicz [53], [54] had shown that a metrizable topological vector space E is locally convex if it is possible to define $\int f\, d\mu$ for all $f \in C(X,E)$, and all measures μ on E, and this integral has reasonable properties.

B. Gramsch [31] and independently D. Przeworska-Rolewicz and S. Rolewicz [64] observed that Żelazko's proofs were possible because the functions which one integrates in Banach algebra theory are limits of rapidly convergent series, and such limits are integrable. This meant that the standard Banach algebra proofs could be used again, and led to a simplification of Zelazko's proofs.

P. Turpin and the author generalized [76] the results above, and obtained essentially the results of paragraph 3, as an application of their theory of integration of differentiable (C_∞) vector valued functions which take their values in a locally pseudo-convex space.

Proposition 16 is due to P. Turpin [73].

CHAPTER VI

THE HOLOMORPHIC FUNCTIONAL CALCULUS

1. The polynomially convex joint spectrum : Let us first fix some notations. If X is compact in $\mathbb{C}^n$, $\mathcal{O}(X)$ is the direct limit of the algebras $\mathcal{O}(U)$, of holomorphic functions on neighbourhoods U of X. We put on $\mathcal{O}(X)$ the direct limit boundedness, a subset $B \subseteq \mathcal{O}(X)$ is bounded if it is possible to find some neighbourhood U of X to which all the elements of B extend, the extensions being uniformly bounded on U.

If $d_1, \ldots, d_n$ are non negative real numbers, $\mathcal{O}_{d_1, \ldots, d_n}$ will be $\mathcal{O}(X)$ where X is the direct product of the closed discs $|z_i| \leq d_i$. This set X is a compact polydisc in $\mathbb{C}^n$.

Let now $a_1, \ldots, a_n$ be regular elements of a commutative b-algebra with unit, let $\rho(a_i) \leq d_i$, and let $F \in \mathcal{O}_{d_1, \ldots, d_n}$ It is quite standard to put

$$F[a_1, \ldots, a_n] = \Sigma F_{r_1, \ldots, r_n} a_1^{r_1} \ldots a_n^{r_n}$$

where

$$F(z_1, \ldots, z_n) = \Sigma F_{r_1 \ldots r_n} z_1^{r_1} \ldots z_n^{r_n}$$

The series converges because $F_{r_1 \ldots r_n} = 0\,(\Pi_i (d_i + \varepsilon)^{-r_i})$, for ε small enough, and $a_i^{r_i} = 0(d_i + \varepsilon/2)^{r_i}$, so that we have a convergent majorant.

The mapping $F \to F[a]$ is a bounded homomorphism of $\mathcal{O}_{d_1 \dots d_n}$ into $\mathfrak{a}$ which maps unit on unit, and z_i on a_i, where z_i is the ith coordinate function.

<u>Definition 1</u> : Let X be a compact subset of $\mathbb{C}^n$. The polynomially convex hull of X is the set

$$\tilde{X} = \{(z_1, \dots, z_n) \mid \forall\ P : |P(z)| \leq \max_{x \in X} |P(x)|\}$$

where P ranges over the set of polynomials on $\mathbb{C}^n$ with complex coefficients.

The polynomially convex joint spectrum of $(a_1, \dots, a_n)$ will be the polynomially convex hull of the joint spectrum. We shall call this set $\tilde{sp}_r(a_1, \dots, a_n)$. If P is polynomial, we know that

$$sp_r\ P(a) = P\ (sp_r\ (a))$$

and

$$\rho(P\ (sp_r\ a)) = \max_{t \in P(sp_r\ a)} |t|$$

Hence :

<u>Proposition 1</u> : We have

$$\tilde{sp}_r\ (a_1, \dots, a_n) = \{(z_1, \dots, z_n) \in \mathbb{C}^n \mid \forall\ P : |P(z)| \leq \rho(P(a))\}$$

when $a_1, \dots, a_n$ are regular elements of $\mathfrak{a}$.

We shall prove the following result in this paragraph. This proposition will be superseded later (proposition 4) :

<u>Proposition 2</u> : It is possible to construct a bounded morphism

$$\mathcal{O}(\widetilde{sp}_r\ (a_1, \ldots, a_n)) \to \mathfrak{a}$$

which maps unit on unit, and z_i on a_i, where z_i is the ith coordinate function on $\mathbb{C}^n$.

We start out with a function $f \in \mathcal{O}(\widetilde{sp}_r\ (a_1, \ldots, a_n))$ and consider an open neighbourhood U to which f extends. We still call f this extension to U.

By a compactness argument, it is possible to find $P_1, \ldots, P_N$, a finite number of polynomials, in such a way that $(z_1, \ldots, z_n) \in U$ as soon as

$$|z_i| \leqslant \rho(a_i),\ (i=1, \ldots, n),\ \text{and}\ |P_k(z)| \leqslant \rho(P_k(a)),\ (k=1, \ldots, N).$$

We let $d_i = \rho(a_i)$, $e_k = \rho(P_k(a))$, consider the algebra

$$\mathcal{O}_{d,e} = \mathcal{O}_{d_1 \ldots d_n\, e_1 \ldots e_n} \quad \text{and} \quad \mathcal{O}(X) \quad \text{where}$$

$$X = \{(z_1, \ldots, z_n) \in \mathbb{C}^n \mid \forall\ i : |z_i| \leqslant \rho(a_i),\ \forall\ k : |P_k(z)| \leqslant e_k\}$$

so that X is compact, contained in U, and $F(z,y) \to F(z, P(z))$ is a morphism $\mathcal{O}_{d,\ e} \to \mathcal{O}(X)$.

<u>Fact 1</u> : The mapping $F(z,y) \to F(z, P(z))$ is a surjective morphism. Its kernel is generated by the polynomials $y_k - P_k(z)$, $k=1, \ldots, N$.

The fact that the mapping is surjective and that its kernel is generated by the functions $y_k - P_k(z)$ follows from the Oka-Cartan ([62], [14]) theory of ideals of holomorphic functions.

We observe that $\mathcal{O}_{d, e}$ is a b-space with a countable boundedness. Its image by the morphism $\mathcal{O}_{d, e} \to \mathcal{O}(X)$ has a completant bounded structure of countable type. This completant bounded structure is maximal on $\mathcal{O}(X)$. But the usual bounded structure of $\mathcal{O}(X)$ is completant, and larger. So the two boundednesses coincide. (We apply Buchwalter's closed graph theorem, chapter II, proposition 12, corollary 3). This yields

Fact 2 : The b-algebra $\mathcal{O}(X)$ is naturally isomorphic with the quotient of $\mathcal{O}_{d, e}$ by the ideal generated by the polynomials $y_k - P_k(z)$, $k=1, \ldots, N$.

We consider now the morphism $F(z,y) \to F[a, P(a)]$, $\mathcal{O}_{d, e} \to \mathcal{a}$. This induces a morphism $\mathcal{O}(X) \to \mathcal{a}$. The image of $f \in \mathcal{O}(X)$ by this quotient morphism will be called $f[a]$.

Proposition 2 is now morally proved. We start with $f \in \mathcal{O}(\tilde{sp}\ a)$, we extend f to some neighbourhood U of $\tilde{sp}\ a$, we choose polynomials P_k, $(k=1, \ldots, N)$ in such a way that $X \subseteq U$. Letting f_1 be the extension of f to U, then $f_2 \in \mathcal{O}(X)$ the restriction of f_1, we can define $f_2[a]$. If B had been a bounded subset of $\mathcal{O}(\tilde{sp}\ a)$, we could have found U in such a way that the elements of B have uniformly bounded extensions to U. The restriction to X of these extensions are then bounded. The set of $f_2[a]$, $f_2 \in B$ that we obtain in this way is then bounded.

To end the proof, we must show that $f_2[a]$ does not depend on the choice of $P_1, \ldots, P_N$. We must therefore consider polynomials $Q_1, \ldots, Q_M$ in such a way that $Y \subseteq U$ if

$$Y = \{(z_1, \ldots, z_n) \mid |z_i| \leq d_i, \ |Q_j(z)| \leq e_j\} \text{ where } e_j = \rho(Q_k(a)).$$

We have mappings $\mathcal{O}(U) \to \mathcal{O}(X) \to \mathcal{O}(X \cap Y)$, and $\mathcal{O}(U) \to \mathcal{O}(Y) \to \mathcal{O}(X \cap Y)$, which all are obtained by means of restriction maps; the two compositions $\mathcal{O}(U) \to \mathcal{O}(X \cap Y)$ coincide therefore. As far as the mappings $\mathcal{O}(X \cap Y) \to \mathfrak{a}$ are concerned, we observe that both map z_i on a_i, unit on unit, $P(z)$ on $P(a)$ when P is a polynomial. By Runge's theorem, polynomials are dense in $\mathcal{O}(X \cap Y)$, so that these conditions determine the bounded morphism $\mathcal{O}(X \cap Y) \to \mathfrak{a}$.

This conclude the proof of proposition 2.

2. <u>The Arens-Calderòn trick</u> : Proposition 2 associates an element of $\mathfrak{a}$ to every $(a_1, \ldots, a_n) \in \mathfrak{a}^n$, and every $f \in \mathcal{O}(\widetilde{sp}_r(a_1, \ldots, a_n))$. This element might reasonably be called $f(a)$. We shall call it $f[a]$, the shape of the brackets indicating that we are not considering the value of f at a, but an element constructed in a relatively complicated way from f and a.

We wish to construct $f[a]$ when $f \in \mathcal{O}(sp_r(a_1, \ldots, a_n))$. This will be possible. The device used in this construction is called the "Arens-Calderon trick".

Let $a_1, \ldots, a_n$ be regular elements of $\mathfrak{a}$. Let then $b_1, \ldots, b_N$ be further regular elements. Projection $\mathbb{C}^{n+N} \to \mathbb{C}^n$, $(z,y) \to z$ maps $sp_r(a, b)$ onto $sp_r\, a$, and $\widetilde{sp}_r(a, b)$ into $\widetilde{sp}_r\, a$. This projection will be called Π_b.

<u>Proposition 3</u> : We have

$$sp_r(a) = \bigcap_b \Pi_b \; \widetilde{sp}_r(a, b)$$

where b ranges over all N-uples of regular elements, N arbitrary.

This is trivial. Let $(s_1, \dots, s_n) \notin sp_r\, a$. Choose regular elements $b_1, \dots, b_n$ such that $\Sigma(a_i - s_i)\, b_i = 1$. The equation $\Sigma(x_i - s_i)\, y_i = 1$ is satisfied identically on $sp_r\,(a, b)$. It is a polynomial equation and is therefore satisfied on $\widetilde{sp}_r(a, b)$. This shows that we cannot find $t \in \mathbb{C}^n$ with $(s, t) \in \widetilde{sp}_r(a, b)$, i.e. that $s \notin \Pi_b\, \widetilde{sp}_r(a, b)$.

Proposition 4 : It is possible to construct in a natural way a morphism $\mathcal{O}(sp_r(a_1, \dots, a_n) \to \mathfrak{a}$, which maps z_i on a_i, and unit on unit.

Let U be a neighbourhood of $sp_r(a_1, \dots, a_n)$ and f be holomorphic on U. A compactness argument yields a finite system $b_1, \dots, b_N$ of regular elements of $\mathfrak{a}$, such that $\Pi_b\, \widetilde{sp}_r(a, b) \subseteq U$. Then, $f \circ \Pi_b$ is holomorphic on a neighbourhood of $\widetilde{sp}_r(a, b)$, proposition 2 allows to define $f \circ \Pi_b\, [a, b]$. We would like to put $f\,[a] = f \circ \Pi_b\, [a, b]$, but we can only do so if the element constructed in this way is independent of b.

But it is. If $b'_1, \dots, b'_N$ are further regular elements of $\mathfrak{a}$, we have

$$f \circ \Pi_b\, [a, b] = f \circ \Pi_{b,b'}\, [a, b, b']$$

Just put $g = f \circ \Pi_b \in \mathcal{O}(\widetilde{sp}_r(a, b))$. Then $g \circ \Pi_{b'} = f \circ \Pi_{bb'}$.

The mappings $g \to g\,[a, b]$ and

$$g \to g \circ \Pi_{b'}\, [a, b, b']$$

are bounded homomorphisms $\mathcal{O}\,(\widetilde{sp}_r(a, b) \to \mathfrak{a}$ which map unit on unit z_i on a_i, and y_k on b_k. Runge's theorem shows that such a homomorphism is unique. This proves proposition 4.

Since $sp_r(a)$ does not satisfy any analytic convexity condition, there may be more than one morphism $\mathcal{O}(sp_r(a)) \to \mathfrak{a}$ which maps unit on unit and z_i on a_i.

The morphism constructed above is natural. This is not a very satisfactory characterization.

If $b_1, \dots, b_N$ are further regular elements,

$$f[a] = f \circ \Pi_b [a, b]$$

by construction. We shall see that this, and the fact that the mappings $\mathcal{O}(\mathrm{sp}_r\, a) \to \mathcal{A}$ are b-algebra morphisms, are sufficient to characterize the family of morphisms considered.

Proposition 5 : Assume that a morphism $f \to \phi_a(f)$, $\mathcal{O}(\mathrm{sp}_r\, a) \to \mathcal{A}$, mapping z_i on a_i and unit on unit is defined whenever $a = (a_1, \dots, a_n)$ is a system of regular elements of $\mathcal{A}$. Assume also that

$$\phi_{a,b} \circ \Pi_b = \phi_a$$

when $b = (b_1, \dots, b_N)$ is a further system of regular elements. When this is the case, $\phi_a(f) = f[a]$.

We assume that $f \in \mathcal{O}(\mathrm{sp}_r\, a)$ has an extension f_1 to a neighbourhood U of $\mathrm{sp}_r\, a$, we choose $b = (b_1, \dots, b_N)$ in such a way that $\Pi_b(\widetilde{\mathrm{sp}}_r(a, b) \subseteq U$. We let $g = f_1 \circ \Pi_b \in \mathcal{O}(\widetilde{\mathrm{sp}}_r(a, b))$. Runge's theorem shows that $\phi_{a,b}(g) = g[a, b]$, but $\phi_{a,b}(g) = \phi_a \circ \Pi_b(g) = \phi_a(f)$, while $g[a, b] = f \circ \Pi_b [a, b] = f[a]$. This yields is the required result.

3. $\sigma(\mathcal{A}^*, \mathcal{A})$ -holomorphic functions : Let E be a locally convex space, U an open subset of E. A complex-valued function on U is holomorphic if it is continuous, and if its restriction to every finite dimensional space is holomorphic.

A sheaf is defined in this way on E. The algebra of holomorphic functions on U is called $\mathcal{O}(U)$.

It is not difficult to show that the local algebra $\mathcal{O}_x$ is the direct limit of algebras $\mathcal{O}_{x,\nu}$. We let ν range over the continuous semi-norms of E, E_ν is the quotient of E by the kernel of the semi-norm ν, normed by $\tilde{\nu}$, the norm induced by ν on the quotient, $q_\nu : E \to E_\nu$ is the quotient mapping. $\mathcal{O}_{x,\nu}$ is the ring of holomorphic germs at $x_\nu = q_\nu(x)$ on the normed space E . We embed $\mathcal{O}_{x,\nu}$ in $\mathcal{O}_x$ by the mapping $f_\nu \to f_\nu \circ q$.

The proof of the fact that $\mathcal{O}_x$ is the direct limit of the algebras $\mathcal{O}_{x,\nu}$ uses essentially the observation that f belongs to image of $\mathcal{O}_{x,\nu}$ when f is bounded on $\{y \mid \nu(x - y) < 1\}$.

<u>Proposition 6</u> : Let $X \subseteq E$ be compact. For each continuous seminorm, let $X_\nu = q_\nu(X)$. The ring $\mathcal{O}(X)$ of sections on X of the holomorphic sheaf of E is the direct limit of the rings $\mathcal{O}(X_\nu)$ of sections on X_ν of the holomorphic sheaf on E_ν.

Let u be a section on X of the holomorphic sheaf on E. Let $x \in X$. We find a semi-norm ν and a holomorphic germ u_ν at $x_\nu = q_\nu x$ such that $u = u_\nu \circ q_\nu$. The equality is valid in x, and in all points $y \in X$ of some neighbourhood V of x. We choose ν large enough, so that $V = \{y \mid (y - x) < 1\}$.

Let then $x_1, \dots, x_n$ be a finite number of points chosen in such a way that $W_1, \dots, W_n$ cover X where $W_i = \{y \mid \nu_i(y - x) < \frac{1}{2}\}$.

Let then $x_1, \dots, x_n$ be a finite set of points of X, let $\nu_1, \dots, \nu_n$ be the semi-norms associated to the x_i by the construction above, assume that $X \subseteq W_1 \cup \dots \cup W_n$ where $W_i = \{y \mid \nu_i(y-x) < \frac{1}{2}\}$

Let also $\nu = 2(\nu_1 + \ldots + \nu_n)$. It is clear that u is the image of a section of the holomorphic sheaf of X_ν on E_ν.

Let now $\mathcal{A}$ be a commutative b-algebra. $\mathcal{A}_r$ is the algebra of its regular elements, and $\mathcal{A}_r^+$ its algebraic dual. The set $\mathcal{M}$ of the multiplicative linear forms of $\mathcal{A}_r$ is weakly, $\sigma(\mathcal{A}_r^+, \mathcal{A}_r)$ - compact in $\mathcal{A}_r^+$. We take the holomorphic sheaf on $\mathcal{A}_r^+$ for the topology $\sigma(\mathcal{A}_r^+, \mathcal{A}_r)$.

$\mathcal{O}(\mathcal{M})$ is the direct limit of algebras $\mathcal{O}(\mathcal{M}_\nu)$, where ν ranges over the continuous semi-norms. A continuous semi-norm is characterized up to equivalence by its kernel, this kernel is a weakly closed subspace of finite codimension. This weakly closed subspace is othogonal to a linearly independent set $(a_1, \ldots, a_n)$ in $\mathcal{A}_r$.

And if we follow up the succession of obvious identifications, we see that $\mathcal{M}_\nu$ is identified with $sp_r(a_1, \ldots, a_n)$, while $\mathcal{O}(\mathcal{M}_\nu)$ is just $\mathcal{O}(sp_r(a_1, \ldots, a_n))$.

Consider now all finite linearly independent subsets of $\mathcal{A}_r$, and for each such set $(a_1, \ldots, a_n)$, we let $[a_1, \ldots, a_n]$ be the vector space generated by $(a_1, \ldots, a_n)$, we also consider $sp_r(a_1, \ldots, a_n)$ and $\mathcal{O}(sp_r(a_1, \ldots, a_n))$. It is reasonable to identify $sp_r(a_1, \ldots, a_n)$ with a subset of the dual of $[a_1, \ldots, a_n]$.

If now, $[a_1, \ldots, a_n] \subseteq [b_1, \ldots, b_m]$, restriction induces a (surjective) mapping $sp_r(b_1, \ldots, b_m) \to sp_r(a_1, \ldots, a_n)$; composing with this mapping, we have an imbedding $\mathcal{O}(sp_r(a_1, \ldots, a_n)) \to \mathcal{O}(sp_r(b_1, \ldots, b_m))$.

It is clear - just apply proposition 6 - that $\mathcal{O}(\mathfrak{M})$ is the direct limit of the algebras $\mathcal{O}(sp_r(a_1, \ldots, a_n))$ with the structural mappings $\mathcal{O}(sp_r(a_1, \ldots, a_n)) \to \mathcal{O}(sp_r(b_1, \ldots, b_m))$ described above when $[a_1, \ldots, a_n] \subseteq [b_1, \ldots, b_m]$. We put the direct limit boundedness on $\mathcal{O}(\mathfrak{M})$.

Proposition 7 : A unique bounded homomorphism $\mathcal{O}(\mathfrak{M}) \to \mathfrak{A}_r$ exists, which maps z_a onto a and unit on unit, if $z_a \in \mathcal{O}(\mathfrak{M})$ is the holomorphic function induced by the linear form $t \to \langle a, t \rangle$ on $\mathfrak{A}_r^+$.

This is essentially a rewording of proposition 5.

Note that we may replace $\mathfrak{A}_r^+$ by any subspace of $\mathfrak{A}_r^+$ which is weakly dense in $\mathfrak{A}_r^+$, i.e. by any subspace which separates $\mathfrak{A}_r$.
If $\mathfrak{A}$ is a continuous inverse locally convex algebra, we may introduce $\mathfrak{A}^*$, the space of continuous linear forms on $\mathfrak{A}$ instead of $\mathfrak{A}^+$. This may be a little less unwieldy.

4. Further properties : It is possible to define f[a] with the construction above when f is a scalar-valued holomorphic function. The same construction cannot be applied to algebra-valued functions, because the Oka-Cartan theory of ideals of holomorphic functions does not apply to vector-valued functions, at least not when the functions take their values in an infinite dimensional space.

Fortunately, $\mathcal{O}(X)$ is a nuclear b-space when X is compact, and topological tensor products will make it possible to define f[a] when f is $\mathfrak{A}$-valued, once f[a] has been defined for scalar-valued f. We have

$$\mathcal{O}(X) \hat{\otimes} \mathfrak{A} = \mathcal{O}(X) \hat{\hat{\otimes}} \mathfrak{A} = \mathcal{O}(X, \mathfrak{A})$$

A bounded linear mapping $\mathcal{O}(X) \to \mathcal{A}$ has been constructed. This extends functorially to a mapping of $\mathcal{O}(X) \hat{\otimes} \mathcal{A}$ into $\mathcal{A} \hat{\otimes} \mathcal{A}$. But multiplication in $\mathcal{A}$ define a bounded linear $\mathcal{A} \hat{\otimes} \mathcal{A} \to \mathcal{A}$.

Proposition 8 : Let $(a_1, \ldots, a_n)$ be regular elements of the commutative b-algebra with unit, $\mathcal{A}$. A unique bounded $\mathcal{A}$-linear homomorphism

$$\mathcal{O}(sp(a_1, \ldots, a_n), \mathcal{A}) \to \mathcal{A}$$

can be constructed, which maps f onto $f[a]$ when f is scalar valued.

Let now $\mathcal{A}$ be a locally pseudo-convex algebra with continuous inversion (the inverse is defined on an open set and is continuous there). We assume that the convex boundedness of $\mathcal{A}$ is completant, so that the integrals that we shall need converge to elements of $\mathcal{A}$ rather than to elements of its completion.

Proposition 9 : Every elements of $\mathcal{A}$ is regular. If $U \subseteq \mathbb{C}^n$ is open, the set $U_{\mathcal{A}} \subseteq \mathcal{A}^n$ of $(a_1, \ldots, a_n)$ such that $sp(a_1, \ldots, a_n) \subseteq U$ is an open subset of $\mathcal{A}^n$. For every $f \in \mathcal{O}(U, \mathcal{A})$, the mapping $a \to f[a]$, $U_{\mathcal{A}} \to \mathcal{A}$ is continuous, and Lorch holomorphic.

The elements of $\mathcal{A}$ are regular. Let $x \in \mathcal{A}$, we must show that $(x - s)^{-1}$ exists and is bounded for $|s| > M$, where M depends on x, but

$$(x - s)^{-1} = - s^{-1} (1 - s^{-1} x)^{-1}$$

The right-hand side exists because $(1 - s^{-1}x)$ is near to the unit when s is large. Also, $(1 - t\,x)^{-1}$ is a continuous function of t on its domain, so $(1 - s^{-1}x)^{-1}$ is bounded when s ranges on a neighbourhood of infinity.

The fact that $U_{\mathfrak{A}}$ is open follows from the continuity of the spectral semi-norm ρ. If $(a_1,\ldots,a_n) \in U_{\mathfrak{A}}$, if ε is chosen small enough so that the ε-neighbourhood of the spectrum is contained in $U_{\mathfrak{A}}$, if $h_1,\ldots,h_n$ are chosen small enough, i.e. $\rho(h_1)+\ldots+\rho(h_n)<\varepsilon$, we see that

$$sp(a_1+h_1,\ldots,a_n+h_n)=\{(\hat{a}_1(m)+\hat{h}_1(m),\ldots,\bar{a}_n(\bar{m})+\hat{h}_n(m)\,|\,m\in\mathfrak{M}\}$$

is contained in U.

We prove the continuity of ρ, observing that $\rho(x)<1$ when $x\in V$, if V is balanced neighbourhood of the origin in $\mathfrak{A}$ such that $1 + x$ has an inverse when $x\in V$. If that is the case, x-s has an inverse for all s such that $[s] \geq 1$, the spectrum of x is contained in the open unit disc.

To prove the continuity of $f[a\ ]$ in the neighbourhood of a°, we observe that it is possible, locally, to choose a finite number of elements $b_1^\circ,\ldots,b_m^\circ$, and of polynomials of n+N variables, $P_1(z,y),\ldots$ $P_r(z,y)$, in such a way that

$$f[a] = F[a,b^\circ,P[a,b^\circ]]$$

for all a belonging to some neighbourhood of a°. The function F is holomorphic on a polydisc $|z_i|<\rho(a_i^\circ)+\varepsilon, |y_i|<\rho(b_i^\circ) + \varepsilon, |u_i|<\rho(P_i|a^\circ,b^\circ|) +\varepsilon$.

This observation shows that it is sufficient to consider the case where the open set U is a polydisc. But then

$$f[a_1,\ldots,a_n]=\frac{1}{(2\pi i)^n}\int_{j_1\times\ldots\times j_n} f(z_1,\ldots,z_n)(z_1-a_1)^{-1}\ldots(z_n-a_n)^{-1}dz_1\ldots dz_n$$

The mappings $z_k \to (z_k-a_k)^{-1}$ belong to $C_\infty(j_k,\mathfrak{A})$, and depend continuously on a_k in the topological vector space $C_\infty(j_k,\mathfrak{A})$. The integral depends continuously on the integrand, so $f[a]$ depends continuously on a. Also,

$$(z_k-a_k)^{-1}=\sum_{r=0}^{\infty}(z_k-a_k^o)^{-r-1}(a_k^o-a)^r$$

when a_k is near to a_k^o, the convergence is valid in the vector space topology of $C_\infty(j_k,\mathfrak{A})$. Substituting, we see that the Taylor's formula holds locally

$$f[a]=\Sigma\, f^{(p)}[a^o]\,\frac{(a-a^o)^p}{p\,!}$$

This shows that f is Lorch holomorphic on $\mathfrak{A}$.

4. Some applications : The Šilov idempotent theorem is the first interesting application of the result above. The idempotents of a commutative algebra with unit form a Boolan algebra with the lattice operations.

$$e_1\vee e_2 = e_1 + e_2 - e_1.e_2 \qquad e_1\wedge e_2$$

The clopen subsets of a topological space form a Boolean algebra with union and intersection as lattice operation. A set is clopen if it is open and closed.

Proposition 10. : The idempotent elements of a commutative b-algebra with unit are regular. The Boolean algebra of idempotents of a commutative b-algebra with unit is isomorphic with that of clopen subsets of the maximal ideal space of the set of its regular elements. The isomorphism maps the idempotent e onto $\{m|\hat{e}(m) = 1\}$.

The idempotent e is regular because

$$(e-s)^{-1}=\frac{e}{1-s}-\frac{1-e}{s}$$

We observe that $sp_r(e) = \{0,1\}$ unless $e = 0$ and $sp\ e = \{0\}$ or $e = 1$ and $sp\ e = \{1\}$. The spectral radius is 1 unless $e = 0$.

If e is idempotent, $\hat{e}$ only takes on the values 0,1.
The set

$$X_e = \{m \mid \hat{e}(m) = 1\}$$

is therefore clopen. The mapping $e \to X_e$ is a homomorphism of the Boolean algebra of idempotent sets into that of clopen sets.

This homomorphism is injective. Assume e and f are idempotents and $X_e = X_f$. Then e-ef is idempotent and X_{e-ef} is empty, i.e. $\hat{e}-\widehat{ef}$ is identically zero, $sp(e-ef) = \{0\}$.
We know this cannot be the case with $e - ef$ idempotent unless $e-ef = 0$. Similarly $f - ef = 0$ so $e = ef = f$.

We must still show that the homomorphism is surjective.
Let $\mathfrak{M} = X_1 \cup X_2$ with X_1, X_2 clopen and disjoint. Let u be defined on a neighbourhood of $X_1 \cup X_2$ by $u = 1$ on a neighbourhood of X_1 and $u = 0$ on a neighbourhood of X_2.
Then u is holomorphic on a neighbourhood of $X_1 \cup X_2$, $u^2 = u$, and the homomorphism constructed in Proposition 5 maps u onto an idempotent e. It is clear that $X_e = X_1$.

<u>Proposition 11</u>. : Let D be the open unit disc, $|z| < 1$, let $F(z_1,\dots,z_n, y)$ be a holomorphic function on D^{n+1}, and let $a_1,\dots,a_n$ be regular elements of $\mathfrak{A}$, a commutative b-algebra with unit. Assume that is possible to find a continuous mapping $\beta : \mathfrak{M} \to D$ with the following properties

a) $F(\hat{a}_1(m),\dots,\hat{a}_n(m), \beta(m)) = 0$

b) the derivative $\partial F/\partial Y$ does not vanish in any point

$$(\hat{a}_1(m),\dots,\hat{a}_n(m),\beta(m)).$$

Under those hypotheses, it is possible to find a unique, regular element b, such that $\hat{b} = \beta$ and $F(a_1,\ldots,a_n,b) = 0$.

We may as well begin by disposing of the uniqueness. We assume that we have found b, with $\hat{b}(m) = \beta(m)$, a solution of the equation $F(a_1,\ldots,a_n,b) = 0$, and consider a complex variable h, putting

$$f(h) = F(a,b+h)$$

The function f is holomorphic and $\mathcal{A}$-valued for small values of h. Also $\frac{\partial f}{\partial h}$ is invertible, because clearly

$$\left(\frac{\partial f}{\partial h}\right)^{\wedge}(m) = \frac{\partial F}{\partial y}(\hat{a}_1(m),\ldots,\hat{a}_n(m),\hat{b}(m))$$

and this does not vanish. A straightforward computation on power series, and standard majorations, yield a series g, convergent on a neighbourhood of the origin, and such that $g_\circ f$ is the identity.

We assume that $F(a_1,\ldots,a_n,b') = 0$ with $\hat{b}'(m) = \hat{b}(m)$ so that $b'=b+\eta$ with η in the radical of the algebra. Also

$$F(a_1,\ldots,a_n,b+\eta) = f(\eta) = 0$$

and $g(f(\eta)) = 0$. But $g(f(\eta)) = g_\circ f(\eta)$ -the composition $g_\circ f$ is the power series obtained by formal composition of power series -but $g_\circ f$ is the identity, i.e. $g_\circ f(\eta) = \eta$.

This shows $\eta = 0$.

The existence proof uses proposition 4. Because $\frac{\partial F}{\partial y}(\hat{a}_1(m),\ldots,\hat{a}_n(m), \beta(m))$ never vanishes, we may use the implicit function theorem, and locally define a holomorphic germ $\tilde{\beta}$, i.e. define β as a holomorphic function of $\hat{a}_1,\ldots,\hat{a}_n$. This gives us a section $\tilde{\beta}$ of the holomorphic sheaf over X.

The homomorphism described in proposition 4 maps $\tilde{\beta}$ onto an element $b \in \mathcal{A}$, which is regular and a solution of the equation.

Proposition 12. : Let $\mathcal{A}$ be a commutative b-algebra with unit and $\mathcal{A}_r$ the algebra of its regular elements with the Allan boundedness. Let $\mathfrak{M}$ be the maximal ideal space of $\mathcal{A}_r$.
The group of components of $GL_n(\mathcal{A}_r)$, the n x n invertible matrices on $\mathcal{A}_r$ is isomorphic to the group of cohomology classes of mappings

$$\mathfrak{M} \to GL_n .$$

Corollary : The group of components of the group of invertible elements of $\mathcal{A}_r$ is isomorphic to $H^1(\mathfrak{M}, Z)$ the first Čech cohomology group of $\mathfrak{M}$ with integer coefficients.

It is not a priori clear what is meant by "the group of components" of $GL_n(\mathcal{A}_r)$ since $\mathcal{A}_r$ does not have a topology. We shall see that every element of the "component of the identity" can be connected to the identity by a path $I \to GL_n(\mathcal{A}_r)$ which is continuous for the topology of $GL_n(\mathcal{A}_B)$, B a suitably chosen convex completant, idempotent set. On the other hand, the "component of the identity" in $GL_n(\mathcal{A}_r)$ is clopen for the topology determined on $GL_n(\mathcal{A}_r)$ by the spectral radius semi-norm on $\mathcal{A}_r$. This leaves a broad choice of topologies and convergence structures for which proposition 7 is valid.

Homotopy classes of mappings are multiplied by pointwise multiplication.

The corollary follows from the proposition : we are investigating the components of $GL_1(\mathcal{A}_r)$, this group is isomorphic to the group of homotopy classes of mappings $\mathfrak{M} \to \mathbb{C}_o$. We have the exact sequence of sheaves :

$$0 \to \underset{\sim}{\mathbb{Z}} \to \underset{\sim}{\mathbb{C}} \to \underset{\sim}{\mathbb{C}}_o \to 0$$

where $\underset{\sim}{\mathbb{Z}}$, $\underset{\sim}{\mathbb{C}}$, $\underset{\sim}{\mathbb{C}}_o$ designate the sheaves of continuous mappings into $\mathbb{Z}$, $\mathbb{C}$ and $\mathbb{C}_o$ respectively. A mapping $\mathfrak{M} \to \mathbb{C}_o$ happens to be homotopic to zero if and only if it can be lifted to a mapping $\mathfrak{M} \to \mathbb{C}$, so -we are studying really

$$H_o(\mathfrak{M}, \underset{\sim}{\mathbb{C}}_o) \,/\, \exp H_o(\mathfrak{M}, \underset{\sim}{\mathbb{C}})$$

This is isomorphic to $H_1(\mathfrak{M}, \mathbb{Z})$ because the sheaf $\underset{\sim}{\mathbb{C}}$ is fine so $H_1(\mathfrak{M}, \underset{\sim}{\mathbb{C}}) = 0$.

The proof of proposition 8 will follow from the results :

a) A continuous mapping $\mathfrak{M} \to GL_n$ is always homotopic to an analytic mapping $\mathfrak{M} \to GL_n$.

b) If an analytic mapping $f : \mathfrak{M} \to GL_n$ is homotopic to a constant, there is a homotopy $\phi : \mathfrak{M} \times I \to GL_n$ such that $\phi_o = e$, a constant, $\phi_1 = f$, and $t \in I$, $\phi_t : \mathfrak{M} \to GL_n$ is analytic.

These results have been proved by Grauert [35], [36] in the finite dimensional case. They are still applicable here, as a straightforward application of the Arens-Calderón trick shows.

Notes and remarks : The theory developped in paragraph 1 and 2 is due to Šilov [71], the author [79], and Arens and Calderon [9]. The author's work was independent from Šilov's, Arens and Calderón's from the author's.

Šilov, Arens, and Calderón constructed the holomorphic functional claculus with the Weil integral formula [90]. This formula is unwieldy, at least in the present setting. They dit not prove that the mapping $f \to f[a]$ was a homomorphism, but they dit show that $f[a]\hat{}(m) = f(\hat{a}(m))$. When $\mathfrak{a}$ is semi-simple, it follows that $f \to f[a]$ is a homorphism. Also, many applications of the holomorphic functional calculus follow already from the results obtained.

The construction given here, with the Oka-Cartan theorems, is due to the author (loc. cit.). The author was not sure at the time that the Arens-Calderón trick was "moral", so that the relegated it to an appendix. It was Arens and Calderón who pointed out the significance of the device.

Šilov, Arens, and Calderón were studying Banach algebras. The author's aim was -already- to develop some sort of a theory of locally convex algebras. And he considered the regular elements of a locally convex algebra, this being a first step in the good direction.

We now know that this first step is a relatively trivial generalization. Graham Allan has shown how the algebra of regular elements was in a natural way a direct limit of Banach algebras. The regular spectrum of $(a_1,\ldots,a_n)$ in the algebra $\mathfrak{a}_B$ where B ranges over the bounded, closed, absolutely convex, idempotent sets.

In most of the developments of this chapter , we could have effected the construction in one of the algebra $\mathfrak{a}_B$, and then obtained the desired result by a compactness argument.

Paragraph 3 is a straightforward extension of the analytic functional calculus. There is hardly any reason to call the analytic functional calculus "infinite dimensional" since each of the functions depends only on a finite number of variables. It is L. Schwartz who attracted the author's attention to proposition 6, to the fact that an algebra of holomorphic functions which it is natural to consider in this context is the algebra of sections of an analytic sheaf of a locally convex space.

Proposition 10 is due to Šilov [71] at least when $\mathcal{A}$ is a finitely generated Banach algebra. Proposition 11 is due to Arens and Calderón [9]. The main purpose of Arens and Calderón's paper was the extension of Šilov's results to infinitely generated Banach algebras. Šilov felt that this extension would require new techniques in functional analysis and it did require the Arens-Calderón trick. The fact must be mentioned that the Šilov idempotent theorem is used more often to prove that a maximal ideal space is connected than to find idempotents in an algebra whose maximal ideal space is known to be disconnected.

The corollary of proposition 12 is the Arens-Royden theorem [7]. Proposition 12 itself is due to Arens [8]. It is part of a program, of investigating the properties of a Banach algebra with depend only on the topological structure of the maximal ideal space.

Another construction of the holomorphic functional calculus can be given. This uses an integral formula due to Fantappié [50]. Some results, involving non-regular elements of a b-algebra can be proved with this method [81]. A simplified version, sufficient to obtain the holomorphic functional calculus in Banach algebra theory has appeared [11].

J.P. Ferrier [21], [22], [23] and I.Cnop [15] have recently shown that the unsimplified version could be applied to obtain results about analytic functions whose growth at the boundary of the domain is limited in a reasonable way.

CHAPTER VII

Fréchet algebras

The reader will not find here a personal exposition of the subject, nor will he find any historical notes, the author has not studied the theory as much as he could have.

Some results must be mentioned however. The reader will find them in W. Zelazko [99] , paragraph 7, and A. Wilanski [93] , chapter I, paragraph 5.

Proposition 1 : Let A be an algebra , τ a Fréchet topology on A, assume that multiplication is separately continuous. Multiplication is then jointly continuous.

We shall show that u : E x F → G is jointly continuous if E is a Baire topological vector space, F a metrizable topological vector space G a topological vector space, and u a separetely continuous bilinear mapping.

We let U be any neighbourhood of the origin in G, and let V be a closed neighbourhood of the origin such that U V - V. We also let $W_1,\ldots,W_n,\ldots$ be a fundamental sequence of neighbourhoods of the origin in F and let

$$A_n = \{x\in E \mid \forall\ y\in W_n : u(x,y)\in V\}$$

Because u is separately continuous , A_n is closed , $\bigcup A_n = E$. But E is a Baire space, the interior of A_n is not empty for n large enough, $A_n - A_n = B_n$ is a neighbourhood of the origin, and

$$u(B_n,W_n) = u(A_n,W_n) - u(A_n,W_n) \subseteq V - V \subseteq U .$$

This proves the required continuity property.

Proposition 1 makes it clear what a Fréchet algebra should be.

Proposition 2. : Let A be a Fréchet algebra, V the set of its invertible elements. The mapping $x \to x^{-1}$, $V \to V$ is continuous if and only if V is a G_δ set.

d will be distance which defines the topology of A.
Assume that $x \to x^{-1}$ is continuous on V. We shall say that $x \in W_\varepsilon$ if $x \in \bar{V}$, and if futhermore some $\eta > 0$ can be found such that $d(y^{-1} , z^{-1}) < \varepsilon$ as soon as $y \in V$, $z \in V$, $d(x,y) < \eta$, $d(x,z) < \eta$.

The set W_ε is an open subset of the closed $\bar{V}$. It contains V because x^{-1} depends continuously on x.

The intersection of the sets W_ε is a (relative) G_δ-set in the closed set $\bar{V}$. This is therefore a G_δ-set in A. We shall show that $V = \cap W_\varepsilon$.

It is clear that $V \subseteq \cap W_\varepsilon$. Let then $x \in \cap W_\varepsilon$. Let $y_n \in V$, $y_n \to x$. Then $d(y_n^{-1} , y_{n'}^{-1}) \to 0$ as n, n' both tend to infinity. So $y_n^{-1} \to z$ as $n \to \infty$, and $xz = \lim y_n . y_n^{-1} = 1$.
This proves half the theorem.

If V is a G_δ set, it is possible to find on V a complete metric ρ which defines on V the induced topology. The conclusion follows therefore from the :

<u>Lemma</u> : Let G be a group, and also a complete metric space with the metric ρ . Assume that multiplication is separately continuous. Inversion is then continuous on G.

We shall show that every sequence $x_n \to e$(the unit) has a subsequence $y_k = x_{n_k}$ such that $y_k^{-1} \to e$.

The sequence y_k will be chosen by induction. Assume that $y_1,\ldots,y_k$ have been chosen. Put $p_k = y_1 \ldots y_k$. It is possible to choose a large enough n_{k+1} in such a way that

$$\rho\,(p_k,p_{k+1}) < 2^{-k-1},\ \rho(p_k y_s^{-1},\ p_{k+1}y_s^{-1}) < 2^{-k-1}$$

for $s = 1,\ldots,k+1$ if we put $y_k = x_{n_k}$, $p_{k+1} = p_k\, y_{k+1}$.

Such a choice of n_{k+1} is possible because $x_n \to e$, because for each value of k, only a finite number of identities must be verified, and each identity is verified for large enough k because multiplication is separately continuous.

We let $p = \lim p_k$, $q_s = \lim p_k y_s^{-1}$. These limits exist because of our construction, and also

$$\rho(p,p_k) < 2^{-k},\ \rho(q_s,p_k y_s^{-1}) < 2^{-k}$$

Furthemore

$$\begin{aligned}\rho(p,q_s) &\le \rho(p,p_s y_s^{-1}) + \rho(q_s,p_s y_s^{-1}\ )\\ &= \rho(p,p_{s-1}) + \rho(q_s,p_s y_s^{-1})\\ &< 2^{-s+1} + 2^{-s} < 2^{-s+2}\end{aligned}$$

so that $q_s \to p$ when $s \to \infty$. We observe that $q_s = p.y_s^{-1}$, by separate continuity of multiplication, and

$$p^{-1}q_s = y_s^{-1} \to p^{-1}\, p = e$$

This ends the proof.

Corollary : A Fréchet algebra A is a continuous inverse algebra if the set of its invertible elements is open.

An open set is a G_δ .

Corollary : A Fréchet topology on a field k is a field topology if multiplication is separately continuous.

We must discuss one last item. There is some indetermination a priori about what is meant by a complete metrizable group or vector space.

A topological group G is complete, metrizable, if it is metrizable, and complete for its left uniform structure. A sequence x_k then converges if $x_{k'}^{-1}\, x_k \to e$(the unit) when k,k' both tend to infinity. It is known that G is then complete for its right uniform structure.

G is completely metrizable it its topology can be defined by a metric with respect to which G is complete. The metric is not assumed right invariant, or left invariant.

Proposition 3 : A completely metrizable commutative group is complete.

Assume that G is not complete. The completion G_1 of G is a group which contains G strictly. The cosets of G are dense in G_1, completely metrizable, and disjoint. The result will follow from the

Lemma 1 : A Hausdorff space cannot be the union of two or more pairwise disjoint, completely metrizable, dense subspaces.

And this result will follow from the :

Lemma 2 : Let X be a complete metric space, and Y a Hausdorff space. An almost open continuous bijection $f : X \to Y$ is bicontinuous.

f is almost open if the closure of fU is a neighbourhood of fx when U is a neighbourhood of x. The proof of lemma 2 is very similar to that of Banach's closed graph theorem. It will not be given in detail here. The crucial point is the proof that $fB(x,2\varepsilon)$ contains the closure of $fB(x, \varepsilon/2)$ when $B(x,r)$ is the ball of radius r and center in x.

To prove Lemma 1 from Lemma 2, we assume that $X = \bigcup_i S_i$ where each S_i is a complete metric space. Let d_i be a complete metric on S_i which determines its topology, assume d_i has been chosen in such a way that $d_i(x,y) \leq 1$ when $x \in S_i$, $y \in S_i$.
We then define a distance d on $\bigcup_i S_i$ in such a way that $d(x,y) = d_i(x,y)$ if x,y both belong to S_i and $d(x,y) = 1$ when $x \in S_i$, $y \in S_j$ with $i \neq j$.

Then (X,d) is a complete metric space : it is the topological sum of the spaces (X_i,d_i). The identity mapping $(X,d) \to (X,\mathcal{T})$ is continuous, bijective, therefore bicontinuous. But none of the sets S_i can be dense in the Hausdorff space $(X, \mathcal{T})$ since these are clopen subsets of (X,d).

Note : A Wilanski states, attempts to prove, and attributes to V.Klee a more general result : a topologically complete group would be complete.

This generalization is not true, at least as the author understands the expression " a complete group".

The error in Wilanski's proof is that he assumes the completion of G to be a group, and it is well known that this is not the case in general. Let I be the unit interval. Let G be the group of homeomorphisms $f : I \to I$ such that $f(0) = 0$, $f(1) = 1$. Let $\mathcal{T}$ be the topology of uniform convergence on G. The right uniform structure of G is then defined by the distance $||u-v||_\infty$, the completion is the set of all continuous, increasing surjections $I \to I$. This is not a group.

G is topological complete. If u,v belong to G, if

$$d(u,v) = ||u-v||_\infty + ||u^{-1} - v^{-1}||_\infty$$

Then d is a complete metric on G.

A topological group has in general a left uniformity $\mathcal{U}_\ell$ and a right uniformity $\mathcal{U}_r$. The upper bound of $\mathcal{U}_\ell$ and $\mathcal{U}_r$ is the two-sided uniformity $\mathcal{U} = \mathcal{U}_\ell \vee \mathcal{U}_r$. The elements of $\mathcal{U}$ are the sets $U_1 \cap U_2$ where $U_1 \in \mathcal{U}_\ell$ $U_2 \in \mathcal{U}_r$.

It is known that a topological group G can be embedded in a group $\tilde{G}$ which is complete for two-sided uniformity. The proof given by Wilansky (and reproduced above without any essential change) shows that :

<u>A completely metrizable group G is complete for its two-sided uniformity</u>

The distance d which we considered above on the group G of homeomorphisms $f : I \to I$ such that $f(0) = 0$, $f(1) = 1$, is precisely the distance which defines two-sided uniformity of G.

CHAPTER VIII

A polarization formula

In this chapter, we shall discuss and give applications of a polarization formula, which yields estimates of $x_1 \dots x_n$ when estimates of x^n are given, and estimates of the polarized forms of the terms of the Taylor expansion of a bounded holomorphic function over a ball in Banach space.

This polarization formula can be used to obtain properties of commutative algebras, with a convex topology or a convex bounded structure. It can also be used to define $F(a)$ when $a \in \mathfrak{a} \hat{\otimes} E$, $\mathfrak{a}$ a Banach algebra, E a Banach space, and F holomorphic on a neighbourhood of the spectrum of a.

The formula will be proved in the first section, a few applications will be given in section 2, 3, 4 in which we obtain estimates of $x_1 \dots x_n$ from estimates of x^n. In sections 5, 6, 7, 8 we shall define the infinite dimensional holomorphic functional calculus.

1. The formula : Let E and F be vector space over fields of characteristic zero, $\hat{u} : E \times \dots \times E \to F$ be a symmetric multilinear mapping, and let $u(x) = \hat{u}(x, \dots, x)$. Polarization is the operation which allows us to recover $\hat{u}$ when u is given.

Proposition 1 : Let $e_1, \dots, e_n$ be elements of E. We have

$$n!\, \hat{u}(e_1, \dots, e_n) = \sum_{I \subseteq \{1, \dots, n\}} (-1)^{n-c(I)} u\Big(\sum_{j \in I} e_j\Big)$$

where $c(I)$ is the number of elements of I.

We shall first prove the formula when $E = F = A$ is a commutative algebra. Once this is done, the reader will have a choice. He may reread the proof and observe that it also yields the result in general. Or he may read further, and find a soft method which shows why Proposition 1 is a corollary of its special case.

$\mathfrak{a}$ is thus a commutative algebra. We consider the multilinear form $\hat{u}(a_1, \ldots, a_n) = a_1 \ldots a_n$. Let

$$w_k = \sum_{c(I)=k} (\sum_{i \in I} a_i)^n$$

We want to show that

$$(-1)^n \, n! \, a_1 \ldots a_n = \sum_0^n (-1)^k \, w_k$$

Consider any term $a_{i_1}^{p_1} \ldots a_{i_r}^{p_r}$ in the development of w_k. We take $i_1 < i_2 < \ldots < i_r$, also $p_1 + \ldots + p_r = n$. This term occurs with the coefficient

$$\frac{n!}{p_1! \ldots p_r!} \binom{n-r}{k-r}$$

since $\binom{n-r}{k-r}$ of the subsets I of $\{1, \ldots, n\}$ which have k elements contain $i_1, \ldots, i_r$, and each of these subsets yields a term similar to the one we consider, with the coefficient $n!/(p_1! \ldots p_r!)$.

The coefficient of $a_{i_1}^{p_1} \ldots a_{i_r}^{p_r}$ in $\sum(-1)^k \, w_k$ is equal to

$$\frac{n!}{p_1! \ldots p_r!} \sum_k (-1)^k \binom{n-r}{k-r}$$

This contains the factor $(1-1)^{n-r}$. It vanishes for $n \neq r$. For $n = r$, it is equal to $(-1)^n \, n!$.

We may apply the above proof when $\mathfrak{a}$ is the algebra of polynomials in n indeterminates $a_1, \ldots, a_n$. We obtain a relation

$$(-1)^n \, n! \, a_1 \ldots a_n = \sum_k (-1)^k \sum_{c(I)=k} (\sum_{i \in I} a_i)^n$$

in the space $\mathcal{P}_n$ of homogeneous polynomials in $(a_1, \ldots, a_n)$.

Let now $\hat{u} : E \times \ldots \times E \to F$ be a symmetric multilinear mapping, and $e_1, \ldots, e_n$ be elements of E. We obtain a linear mapping $\mathcal{P}_n \to F$ which maps $a_{i_1} \ldots a_{i_n}$ onto $\hat{u}(a_{i_1}, \ldots, a_{i_n})$.

This mapping maps the above relation in $\mathcal{P}_n$ onto a relation of F. It is easy to see that this is the required one.

This formula allows us to obtain convex estimates of $\hat{u}$ when convex estimates of u are assumed.

Corollary : Let E, F be real or complex vector spaces. Let $X \subseteq E$, $Y \subseteq F$ be absolutely convex. Let $u_\alpha : E \to F$ $(\alpha \in A)$ be homogeneous polynomial mappings, u_α of degree n_α, and assume that $u_\alpha X \subseteq M_\alpha Y$ for all α. Then

$$\hat{u}_\alpha(X, \ldots, X) \subseteq (2e)^{n_\alpha} M_\alpha Y$$

for all α.

This is not difficult. Let $I \subseteq \{1, \ldots, n_\alpha\}$, then $c(I) \leq n_\alpha$, $\sum_{i \in I} e_i \in n_\alpha X$ if each $e_i \in X$, and

$$u(\sum_{i \in I} e_i) \in M_\alpha n_\alpha^{n_\alpha} Y$$

Now, $n_\alpha! \hat{u}(e_1, \ldots, e_n)$ is a linear combination of 2^{n_α} such terms

$$\hat{u}_\alpha(e_1, \ldots, e_{n_\alpha}) \in M_\alpha \frac{(2n_\alpha)^{n_\alpha}}{n_\alpha!} Y$$

Stirling's formula shows that $n_\alpha! \geq (n_\alpha/e)^{n_\alpha}$, i.e.

$$\hat{u}_\alpha(e_1, \ldots, e_{n_\alpha}) \in (2e)^{n_\alpha} M_\alpha Y$$

2. Fréchet algebras on which entire functions operate : Let $\mathfrak{a}$ be a topological algebra, $a \in \mathfrak{a}$. Let $f = \sum f_n z^n$ be an entire function. It is reasonable to say that f operates on a if $\sum f_n a^n$ converges. It is even more reasonable to say that entire functions operate on $\mathfrak{a}$ if $\sum f_n a^n$ converges each time $\{f_n\}$ is the sequence of Taylor coefficients of an entire function and each time $a \in \mathfrak{a}$.

It is clear that entire functions operate on $\mathfrak{a}$ if $\mathfrak{a}$ is complete, locally m-convex in the sense of Michael [55], i.e. if $\mathfrak{a}$ is an inverse limit of Banach algebras. The following result of

Mitjagin, Rolewicz, and Zelazko [58] is a partial converse.

Proposition 2 : Let $\mathfrak{a}$ be a commutative, complete, metrizable algebra on which analytic functions operate. Then $\mathfrak{a}$ is locally m-convex.

The proof goes in two steps. We first prove the following lemma by a double category argument :

Lemma 1 : Let $\mathfrak{a}$ be a complete metrizable locally convex algebra. Assume that $\lambda_n a^n \to 0$ for every sequence of positive scalars λ_n such that $\lambda_n^{1/n} \to 0$. To each neighbourhood V of the origin in $\mathfrak{a}$ we can associate a neighbourhood W such that

$$V \supseteq \{x^n \mid x \in W, n \in \mathbb{N}, n \neq 0\}$$

We start out with a closed convex balanced neighbourhood U of the origin and a sequence $\{\lambda_n\}$ such that $\lambda_n^{1/n}$ is decreasing and tends to zero. Let V be the set of $a \in \mathfrak{a}$ such that $\lambda_n a^n \in U$ for all n. Then V is closed and balanced. V is absorbing. Consider any $a \in \mathfrak{a}$, since $\lambda_n a^n \to 0$, $\lambda_n a^n \in U$ for all but a finite number of values of n. For those n, $\lambda_n a^n \in \varepsilon^{-1} U$ for $\varepsilon > 0$, small. We take $\varepsilon < 1$, $\lambda_n (\varepsilon a)^n \in U$ for all n. So V has a non empty interior.
Let U_1 be a new neighbourhood of the origin such that $U \supseteq U_1.U_1$.
Let V_1 be the set of $a \in \mathfrak{a}$ such that $\lambda_n a^n \in U_1$ for all $n \geq 1$.
Then V_1 has a non empty interior.
Let $x = 1/2(a + b)$ with $a \in V_1$, $b \in V_1$, then

$$\lambda_n x^n = 2^{-n} \sum \binom{n}{r} \frac{\lambda_n}{\lambda_r \cdot \lambda_{n-r}} \lambda_r a^r \lambda_{n-r} b^{n-r}$$

This belongs to U because $\lambda_r a^r . \lambda_{n-r} b^{n-r} \in U$, $\lambda_n \leq \lambda_r \lambda_{n-r}$ (because $\lambda_n^{1/n}$ is decreasing), and

$$2^{-n} \sum \binom{n}{r} \frac{\lambda_n}{\lambda_r \cdot \lambda_{n-r}} \leq 1$$

So $V \supseteq 1/2(V_1 + V_1)$ is a neighbourhood of the origin.

This is half the proof of the lemma. We observe that we can associate to every sequence of complex numbers c_n such that $|c_n|^{1/n} \to 0$ a sequence of positive reals λ_n such that $|c_n| \leq \lambda_n$,

and $\lambda_n^{1/n}$ decreases to zero.
So, for every such sequence c_n, and every neighbourhood U of the origin, we can find a neighbourhood V of the origin such that $c_n x_n \in U$ each time $x \in V$.

To prove the second half, we consider a neighbourhood U of the origin, take U closed and convex, and then a basis $V_1, \ldots, V_k, \ldots$ of neighbourhoods of the origin in $\mathfrak{a}$.
We let X_k be the set of sequences of scalars c_n, with $|c_n|^{1/n} \to 0$, and $\forall x \in V_k : c_n x^n \in U$. Then x_k is absolutely convex, closed in the space of entire functions, and $\bigcup X_k$ is the space of all entire functions. Baire's theorem shows that X_k must have an interior for large k.

This gives constants M_1, M_2, and a neighbourhood V_k of the origin, such that

$$x^n \in M_1 . M_2{}^n \, U$$

whenever $x \in V_k$. And $x^n \in U$ if $x \in V_k / M_1 . M_2$. The lemma is proved.

To complete the proof, we shall use a rewording of the corollary of proposition 1 :

<u>Lemma 2</u> : Let U and V be absolutely convex subsets of a commutative algebra. Assume that $x^n \in U$ as soon as $x \in V$, $n=1, 2, \ldots$. An absolutely convex V_1 can then be found, with $V_1{}^2 \subseteq V_1$, and $U \supseteq V_1 \supseteq (2\,e)^{-1}\, V$.

V_1 is the absolutely convex hull of the set of $x_1, \ldots, x_n$, with $\forall n : x_i \in V/2e$, and $n \geq 1$. The corollary gives the result.

Combining lemmas 1 and 2, we see that each neighbourhood U of the origin in $\mathfrak{a}$ contains a neighbourhood V, which is absolutely convex and idempotent. The algebra is locally multiplicatively convex.

3. <u>Continuous inverse locally convex algebras</u>. The following result is proved by P. Turpin [73].

Proposition 3 : A commutative, locally convex, continuous inverse algebra is locally multiplicatively convex.

It is possible to associate to each neighbourhood U of the origin in $\mathfrak{a}$ a neighbourhood V in such a way that

$$U \supseteq \{x^n \mid n \in \mathbb{N},\ n \neq 0,\ x \in U\}$$

Once this has been shown, we observe that U and V may be taken absolutely convex, lemma 2 shows that an absolutely convex, idempotent V_1 can be found in such a way that $U \supseteq V_1 \supseteq (2e)^{-1} V$, and proposition 3 follows.

So we must show that $x^n \to 0$ uniformly as $x \to 0$ and $n \in \mathbb{N}$, $n \neq 0$. We choose a neighbourhood of the origin, U_1, such that $\rho(x) < 1/2$ when $x \in U_1$. Then, by the holomorphic functional calculus

$$x^n = \frac{1}{2\pi} \int_0^{2\pi} e^{int} (1 - e^{it} x)^{-1} \, dt$$

If $x \to 0$, $(1 - e^{it} x)^{-1} \to 1$ uniformly, $t \in \mathbb{R}$, and $x^n \to 0$ uniformly, $n \in \mathbb{N}$, $n \neq 0$.

4. The equiregular boundedness : Let now $\mathfrak{a}$ be a commutative b-algebra with unit. An element $a \in \mathfrak{a}$ is regular if $(a - s)^{-1}$ is defined and bounded for s large.

Definition 1 : A set B of regular elements of $\mathfrak{a}$ is equiregular if it is possible to find $M > 0$ in such a way that $a - s$ has an inverse when $a \in B$, $|s| \geq M$, the set

$$\{(a - s)^{-1} \mid a \in B,\ |s| \geq M\}$$

being a bounded subset of $\mathfrak{a}$.

We remember that a is regular if, and only if a^n/M^n is a bounded sequence for M large.

Proposition 4 : B is an equiregular set if and only if it is possible to find some M such that

$$\left\{ \frac{b^n}{M^n} \;\middle|\; b \in B,\ n \in \mathbb{N} \right\}$$

is a bounded set.

Corollary : An equiregular set is bounded.

Assume B equiregular, and assume $(a - s)^{-1}$ defined and bounded for $|s| \geq M$, and $a \in B$. Then, for all $a \in B$, and all n

$$a^n = \frac{M^n}{2\pi} \int_0^{2\pi} e^{nit} (1 - M e^{-it} a)^{-1} dt$$

and $\{a^n/M^n\}$ considered is bounded. If conversely, this set is bounded, and $|s| \geq 2M$, then $-\sum s^{-n-1} a^n = (a - s)^{-1}$ belongs to the completant hull of this bounded set. It follows that B is an equiregular set.

The set $\mathfrak{a}_r$ of regular elements of a commutative b-algebra $\mathfrak{a}$ is a subalgebra of $\mathfrak{a}$.

Proposition 5 : The set of equiregular subsets of $\mathfrak{a}_r$ is an algebra boundedness on $\mathfrak{a}_r$. If B is equiregular, and if B_1 is a completant bounded subset of $\mathfrak{a}$, then $\bigcap_{\varepsilon > 0} (B + \varepsilon B_1)$ is equiregular.

An analysis of the proof of proposition 12, chapter V yields the uniform majorants that we need to prove the first part of proposition 5. It is not more difficult to prove that the sum, or the product of equiregular sets is equiregular, than to show that the sum, or the product of regular elements is regular.

Let now B be equiregular. Let $x \in \bigcap (B + \varepsilon B_1)$ with B_1 completant, $B_1 \supseteq B$. We choose M real, and B_2 bounded in such a way that $(a - s)^{-1} \in B_2$ when $|s| \geq M$, $a \in B$.

Choose $a_n \in B$, $a_n - x \in 2^{-n} B_1$. Then

$$(a_n - s)^{-1} - (a_m - s)^{-1} = (a_m - a_n)(a_n - s)^{-1}(a_m - s)^{-1} \in (2^{-m} + 2^{-n}) B_1 B_2^2$$

This shows that $(a_n - s)^{-1}$ is a Cauchy sequence of the Banach space $\mathfrak{a}_{B_3}$ where B_3 is completant, $B_3 \supseteq B_1 B_2^2$. This sequence has a limit, which is an inverse of $(x - s)$. This limit belongs to the closure of B_2 in the Banach space $\mathfrak{a}_{B_3}$. This completes the proof of proposition 5.

The author does not know whether the equiregular boundedness of a b-algebra is convex. He does not believe it is in general, but does not possess any counterexample either. A convex boundedness can be associated to every vector space boundedness, its elements are the sets with a bounded convex hull.

<u>Proposition 6</u> : The convex boundedness associated to the equiregular boundedness is the Allan boundedness of the algebra of regular elements.

This is a straightforward application of Lemma 2. Let B be equiregular and convex. Choose M large enough, then B_1 convex, bounded, and such that $x^n/M^n \in B_1$ when $x \in B_1$, $n \in \mathbb{N}$. Let B_2 be the absolutely convex hull of

$$\{(x_1 \dots x_n)/(2\ Me)^n \mid n \in \mathbb{N},\ \forall i : x_i \in B\}$$

Then B_2 is bounded, absolutely convex, and idempotent, and $B \subseteq M\ B_2$.

It is clear that the bounded, absolutely convex idempotent sets are equiregular, so the Allan boundedness is the convex boundedness associated to the equiregular one.

5. <u>Restatement of the classical holomorphic functional calculus</u>.
Before giving a statement of the infinite dimensional holomorphic functional calculus, it seems a good idea to reword the main theorems of chapter 6. With this rewording, it will be clear in what way our infinite dimensional statement is a generalization.

The reworded results will not be proved here. The reader may identify an n-dimensional space E with $\mathbb{C}^n$, and $\mathfrak{a} \otimes E$ with $\mathfrak{a}^n$. Once that is done, the corresponcence between the results of chapter 6 and the statements here will be limpid.

Let E be a finite dimensional space, and $\mathfrak{a}$ a commutative Banach algebra with unit. Let $a \in \mathfrak{a} \otimes E$. If m is a maximal ideal, and χ_m the associated multiplicative linear form, a linear $\mathfrak{a} \otimes E \to E$ is associated linear mapping. The image of a is called $\hat{a}(m)$, the set of $\hat{a}(m)$, $m \in \mathfrak{M}$, where $\mathfrak{M}$ is the set of maximal ideals of $\mathfrak{a}$, is the spectrum of a :

$$\text{sp } a = \{\hat{a}(m) \mid m \in \mathfrak{M}\}$$

This is a compact subset of E.

Let now $u \in E^*$. A mapping $\mathfrak{a} \otimes E \to \mathfrak{a}$ is associated to u. The image of u will be called u[a].

Proposition A : A bounded homomorphism $\mathcal{O}(\text{sp } a) \to \mathfrak{a}$ can be found, which maps $u \in E^*$ onto u[a]. The image of $f \in \mathcal{O}(\text{sp } a)$ by this bounded homomorphism is called f[a].

Let $E \subseteq E'$ be two finite dimensional spaces, $\mathfrak{a} \otimes E \subseteq \mathfrak{a} \otimes E'$.
If $a \in \mathfrak{a} \otimes E$, sp $a \subseteq E$. We will have to distinguish the algebra $\mathcal{O}_E(\text{sp } a)$ of holomorphic functions on neighbourhoods of sp a in E, and $\mathcal{O}_{E'}(\text{sp } a)$ of holomorphic functions on neighbourhoods of sp a in E'. Restriction to E is a homomorphism of $\mathcal{O}_{E'}(\text{sp } a)$ onto $\mathcal{O}_E(\text{sp } a)$. We shall call $p_{EE'}$ this homomorphism.

Proposition B : When $f \in \mathcal{O}_{E'}(\text{sp } a)$, we have

$$f[a] = p_{EE'}(f)[a]$$

(this is part of the "Arens-Calderon trick").

Assume now that a morphism $\varphi_a : \mathcal{O}(\text{sp } a) \to \mathfrak{a}$ is associated to each $a \in \mathfrak{a} \otimes E$, this for all finite dimensional spaces E. Assume that $\varphi_a(1) = 1$ for all a, and that $\varphi_a(u) \in u[a]$ whenever $u \in E^*$. Assume also that

$$\varphi_a(f) = \varphi_a(p_{EE'}f)$$

when $a \in \mathfrak{a} \otimes E$ and $E \subseteq E'$.

Proposition C : When these several conditions hold, we have, for all $a \in \mathfrak{a} \otimes E$ and all $f \in \mathcal{O}(\text{sp } a)$

$$\varphi_a(f) = f[a]$$

6. <u>Infinite dimensional statements</u>. Now, E will be a Banach space, $\mathfrak{a}$ a Banach algebra, and $a \in \mathfrak{a} \hat{\otimes} E$, the projective tensor product of $\mathfrak{a}$ and E. Then $a = \sum a_k \otimes e_k$ with $\sum ||a_k||.||e_k|| < \infty$.

If $m \in \mathfrak{M}$ is a maximal ideal of $\mathfrak{a}$, we put $\hat{a}(m) = \sum \hat{a}_k(m)e_k$. The mapping $\hat{a}$ is a continuous mapping of $\mathfrak{M}$ into E. We define

$$\text{sp } a = \hat{a}\ (\mathfrak{M})$$

If $X \subseteq E$ is compact, $\mathcal{O}(X)$ will be the direct limit of the rings of holomorphic functions on neighbourhoods of X. A set $B \subseteq \mathcal{O}(X)$ will be bounded if X has a neighbourhood U to which all $f \in B$ extend, the extensions being uniformly bounded on U. Then $\mathcal{O}(X)$ is a b-space of countable type.

It is possible to define a countable convex compactology on $\mathcal{O}(X)$. Just put on the set of functions which extend to U, in such a way that $|f(z)| \leq M$ on U the topology of pointwise convergence on U. We shall not use this compactology, but it would be surprising if this compactology had no bearing on the further study of $\mathcal{O}(X)$.

Let $u \in E^*$. We define $u[a]$ as $u[a] = \sum u(e_n)\ a_n$ when $a = \sum a_n \otimes e_n$. More generally, let u be a homogeneous polynomial of degree k. Let $\hat{u}$ be the polarized form of u, $\hat{u}$ is a symmetric multilinear form. We define $u[a]$, again when $a = \sum a_n \otimes e_n \in \mathfrak{a} \otimes E$ by

$$u[a] = \sum \hat{u}\ (e_{n_1}, \ldots, e_{n_k})\ a_{n_1} \ldots a_{n_k}$$

We intend to prove the following statements in the next few paragraphs.

<u>Proposition 7</u> : It is possible to define in a natural way a bounded homomorphism, $f \to f[a]$, $\mathcal{O}(\text{sp } a) \to \mathfrak{a}$ which maps each homogeneous form u on E onto $u[a]$.

Let E and F be Banach spaces. A linear mapping $T : E \to F$ is "Fredholm" if T has a finite dimensional kernel, and a closed finite codimensional range. If $T : E \to F$ is any linear mapping and

$a \in \mathfrak{a} \hat{\otimes} E$, $a = \sum a_n \otimes e_n$, then $T(a) = \sum a_n \otimes T e_n \in \mathfrak{a} \hat{\otimes} F$.

<u>Proposition 8</u> : For all linear $T : E \to F$, $a \in \mathfrak{a} \hat{\otimes} E$, $f \in \mathcal{O}(\mathrm{sp}\ Ta)$, we have $f \circ T \in \mathcal{O}(\mathrm{sp}\ a)$ and

$$(f \circ T)\,[a] = f\,[Ta].$$

<u>Proposition 9</u> : Assume that a bounded morphism $\varphi_a : \mathcal{O}(\mathrm{sp}\ a) \to \mathfrak{a}$ has been defined whenever E is a Banach space and $a \in \mathfrak{a} \hat{\otimes} E$. Assume also that $\varphi_a(u) = u[a]$ when u is an homogeneous polynomial on E, and that

$$\varphi_a\,(f \circ T) = \varphi_{Ta}(f)$$

whenever T is a Fredholm mapping, and $f \in \mathcal{O}(\mathrm{sp}\ Ta)$. Then

$$\varphi_a(f) = f\,[a]$$

in all cases.

In our construction, we shall use an algebra of holomorphic functions "of bounded type" on an open subset U of the Banach space. These are the holomorphic functions on U which are bounded on the bounded subsets of U whose distance to the complement of U is positive. We topologize the algebra $\mathcal{O}_b(U)$ of holomorphic functions of bounded type on U by uniform convergence on the subsets of U on which these functions must be bounded.
With this topology, $\mathcal{O}_b(U)$ is a Fréchet, locally convex algebra.

7. <u>Taylor expansions</u>. <u>Lemma 3</u> : Let $a \in \mathfrak{a} \hat{\otimes} E$, let $r > 2e\|a\|$, let U be the ball of radius r and center at the origin in E. If $f(z) \in \mathcal{O}_b(U)$ and $f(z) = \sum f_n(z)$ is its Taylor expansion, then $\sum f_n\,[a]$ converges, and putting $f\,[a] = \sum f_n\,[a]$, the mapping $f \to f\,[a]$, $\mathcal{O}_b(U) \to \mathfrak{a}$ is a continuous homomorphism $\mathcal{O}_b(U) \to \mathfrak{a}$.

The statement is homothetically invariant. We may assume that $r > 2e$, $\|a\| < 1$. Some M can then be found such that $\|f_n\| < M(2e)^{-n}$. If $\hat{f}_n$ is the polarized form of f_n, $\|\hat{f}_n\| < M$, and

$$||f_n\ [a]|| \leq ||\sum \hat{f}(e_{k_1}, \ldots, e_{k_n})\, a_{k_1} \ldots a_{k_n}||$$

$$\leq M\,(\sum ||a_k||\ ||e_k||)^n$$

where $a = \sum a_k \hat{\otimes} e_k$. Since $||a|| < 1$, we can find an expression $a = \sum a_k \hat{\otimes} e_k$ with $\sum ||a_k||\ ||e_k|| < 1$.

The convergence of the series $\sum f_n\ [a]$ and the continuity of the mapping $f \to f\ [a]$ follow easily from these considerations.

We can combine proposition A and lemma 3.

Lemma 4 : Let $a = a_1 \oplus a_2 \in \mathfrak{a} \hat{\otimes} (E_1 \oplus E_2)$, where E_1 is finite dimensional and E_2 is a Banach space. Let $U \subseteq E_1 \oplus E_2$ contain the direct product of an open neighbourhood U_1 of sp a_1 and an open ball U_2 of radius r_2 in E_2, where $r_2 > 2e||a_2||$.

We begin by mapping $\mathcal{O}_b(U)$ into $\mathcal{O}_b(U_1 \times U_2)$ by restriction. It is sufficient to consider the case where $U = U_1 \times U_2$. We remember that $\mathcal{O}(U_1) = \mathcal{O}_b(U_1)$ is nuclear, because U_1 is finite dimensional, therefore

$$\mathcal{O}_b(U_1 \times U_2) = \mathcal{O}_b(U_1)\ \hat{\hat{\otimes}}\ \mathcal{O}_b(U_2) = \mathcal{O}_b(U_1)\ \hat{\otimes}\ \mathcal{O}_b(U_2)$$

Proposition A defines a mapping $\mathcal{O}_b(U_1) \to \mathfrak{a}$, lemma 4 a mapping $\mathcal{O}_b(U_2) \to \mathfrak{a}$. Combining these two mappings with multiplication $\mathfrak{a} \times \mathfrak{a} \to \mathfrak{a}$, we find the desired $\mathcal{O}_b(U_1 \times U_2) \to \mathfrak{a}$.

We notice that there can be only one bounded linear morphism $\mathcal{O}_b(U_1 \times U_2) \to \mathfrak{a}$, which maps the homogeneous polynomial u onto $u[a]$ if U_1 is polynomially convex. The given conditions determine the morphism on $\mathcal{O}_b(U_1)\ \hat{\otimes}\ \mathbb{C}$ and $\mathbb{C}\ \hat{\otimes}\ \mathcal{O}_b(U_2)$. These two algebras together generate a dense subalgebra of $\mathcal{O}_b(U_1 \times U_2)$.

8. Proof of the announced results. Let $a \in \mathfrak{a}\ \hat{\otimes}\ E$. Let U be an open neighbourhood of sp a. Then U contains the ε-neighbourhood of sp a_1 for some $\varepsilon > 0$. We shall choose later $\eta > 0$ small enough, so that all our constructions can be carried out.

Let $a_1 \in \mathfrak{a} \otimes E$, $||a - a_1|| < \eta$. Then $a_1 \in \mathfrak{a} \otimes E_1$ where E_1 is a finite dimensional subspace of E. Let U_1 be the η-neighbourhood of $\mathrm{sp}\, a_1$ in E_1 and W a ball in E of radius $\varepsilon - 2\eta$. When $f \in \mathcal{O}_b(U)$, define a function on $U_1 \times W$, by $g(z_1,\zeta) = f(z_1 + \zeta)$. Then, lemma 5 allows us to define $g[a_1, a-a_1]$ as soon as $2e\eta < \varepsilon - 2\eta$.

This is not quite good enough for our purpose. We shall need polynomially convex domains in order to apply Runge's theorem, and polynomially convex domains come out of the "Arens-Calderón trick" (chapter VI, paragraph), which must be interpreted in our rather abstract setting.

It is possible to construct a finite dimensional space F_1, a surjection $v_1 : F_1 \to E_1$, an element $b_1 \in \mathfrak{a} \otimes F_1$, such that $v_1 [b_1] = a_1$, and a polynomially convex neighbourhood V_1 of $\mathrm{sp}\, b_1$ such that $v_1(V_1) \subseteq U_1$.
We define $S_1 f$ on $V_1 \times W$ by the relation

$$S_1 f(z_1,\zeta) = f(v_1 z_1 + \zeta)$$

and would appreciate defining

$$f\,[a] = S_1 f\,[b_1, a - v_1 b_1]$$

Proposition 9 already follows from these observations. Assume that it is possible to define a family of morphisms with the required properties. Then

$$\varphi_a(f) = \varphi_{b_1,\, a - v_1 b_1}(S_1 f) = S_1 f\,[b_1, a - v_1 b_1]$$

the second equality following **from the** fact that $S_1 f$ is defined on $V_1 \times W$ with V_1 polynomially convex.

To prove proposition 10, we must show that $S_1 f\,[b_1, a - v_1 b_1]$ does not depend on the arbitrary choice of b_1, v_1, etc... as long as $||a - v_1 b_1|| < \eta$ for some small enough η.
We choose therefore a finite dimensional space F_2, an element $b_2 \in \mathfrak{a} \otimes F_2$, a linear mapping $v_2 : F_2 \to E$, assume that $||a - v_2 b_2|| < \eta$, and that a polynomially convex neighbourhood V_2 of $\mathrm{sp}\, b_2$ can be found in such a way that $v_2 V_2$ is contained in

the 2η-neighbourhood of sp a. Then $S_2f(z_2,\zeta)$ can be defined on $V_2 \times W$ by

$$S_2f(z_2,\zeta) = f(v_2z_2 + \zeta)$$

We must compare $S_2f\,[a_2, a - v_2b_2]$ and $S_1f\,[a_1, a - v_1a_1]$.

We consider $b_1 \oplus b_2 \in \mathfrak{a} \hat{\otimes} (E_1 \oplus E_2)$. We know that $||v_1b_1 - v_2b_2|| < 2\eta$. A polynomially convex neighbourhood V_3 of sp $(b_1 \oplus b_2)$ can therefore be found, such that $|v_1z_1 - v_2z_2| < 2\eta$ when $z_1 \oplus z_2 \in V_3$.
We can choose V_3 in such a way that the canonical projections of $E_1 \oplus E_2$ into E_1 and E_2 map V_3 into V_1 and V_2.

We also let W' be the ball of radius $\varepsilon - 4\eta$ in E.

Two mappings $T_1 : \mathcal{O}(V_1 \times W) \to \mathcal{O}(V_3 \times W')$, $T_2 : \mathcal{O}(V_2 \times W) \to \mathcal{O}(V_3 \times W')$ can now be defined, putting

$$T_1f(z_1, z_2, \zeta) = f(z_1,\zeta)$$

$$T_2f(z_1, z_2, \zeta) = f(z_2,\zeta - v_1z_1 + v_2z_2)$$

We observe that $T_1 \circ S_1 = T_2 \circ S_2$.

Also

$$f\,[b_1, a - u_1b_1] = T_1f\,[b_1, b_2, a - u_1b_1]$$

and

$$f\,[b_2, a - u_2b_2] = T_2f\,[b_1, b_2, a - u_1b_1]$$

The proofs of both identities use the polynomial convexity of U_1, of U_2, and the fact that we have two bounded morphisms $\mathcal{O}_b(U_1 \times W) \to \mathfrak{a}$ in the first case, $\mathcal{O}_b(U_2 \times W) \to \mathfrak{a}$ in the second case, which coincide on a generating set of polynomials.

Let now $f \in \mathcal{O}(U)$

$$S_1f\left[b_1, a - v_1\,[b_1]\right] = T_1 \circ S_1f\left[b_1, b_2, a - v_1\,[b_1]\right]$$

$$= T_2 \circ S_2f\left[b_1, b_2, a - v_1\,[b_1]\right]$$

$$= S_2 f \left[b_2, a - v_2 \left[b_2\right]\right]$$

and this is the required result. In our computations, we have used the fact that $2e\eta < \varepsilon - 4\eta$, i.e. $\eta < \varepsilon/(2e + 4)$.

Notes and remarks : Let u be an homogeneous form of degree n on a vector space, and $\hat{u}$ its polarized, n-linear form. It is clear that $\hat{u}(x_1, \ldots, x_n)$ is a linear combination of values of u in the points $x_1, \ldots, x_n$ and in linear combinations of these $x_1, \ldots, x_n$. This simple observation shows that the space of symmetric multilinear forms in n variables is isomorphic with the space of homogeneous polynomials of degree n. It also shows how one can obtain evaluations of a symmetric form $\hat{u}$ when evaluations of u are given.

This remark yields results that are generally satisfactory when n is given. When n varies, it shows that a sequence of positive real constants M_n can be found, for example, but it does not give any idea about the behaviour of M_n.

The explicit polarization formula that we have considered shows that we can take $M_n = (2e)^n$ when a convex evaluation of $u(x)$ is given, the argument x ranging over a convex set. The factor $(2e)^n$ is inessential, since it only involves multiplication of the argument by $2e$, a positive constant.

This sort of argument does not generalize when locally convex spaces are replaced by locally p-convex ones, for instance.
The factor $n^n/n!$ is replaced by $n^{n/p}/n!$, and this grows faster than n^{nt} with $t = p^{-1} - 1$.

And Turpin gives an example of a continuous inverse, Fréchet, locally p-convex algebra, whose topology cannot be defined by continuous multiplicative p-semi-norms (Ref. [73]).
This is also a Fréchet locally p-convex algebra on which all analytic functions operate.

The results in paragraph 2 are due to Mitjagin, Rolewicz, and Zelazko [58]. Those of paragraph 3 are due to Turpin [73].
Neither paragraph contains the other since Mitjagin, Rolewicz and Zelazko assume that the algebra is Fréchet, while Turpin assumes

that the inverse is defined on an open set - he also assumes that the inverse is continuous on its domain, but that follows from the fact that the domain is open when the algebra is Fréchet.

The main theorem in paragraph 4 may be folklore rather than earnest mathematics. The author has no idea that this equiregular boundedness looks like, and does not know any algebra whose equiregular boundedness is not convex.

About paragraphs 5 to 8, the author had nearly proved the result a long time ago, but he had at the time stumbled on the fact that he needed bounds for the polarized forms $\hat{a}$ of a homogeneous polynomial a. The whole construction is saved by the fact that we may take $M_n = (2e)^n$.

The proof is technical, the author sees no obvious applications. Such applications may be lacking because the theory of holomorphic functions on Banach spaces is not yet fully understood. In any case, this lack of applications is no excuse for omitting a result involving analytic functions on Banach spaces from a series of lectures in Brasil.

CHAPTER IX.

Topological extensions of the complex field.

1. p-semi-norms on extensions : Let k be a topological field containing $\mathbb{C}$ as a subfield and inducing on $\mathbb{C}$ its usual topology. Then k is a topological extension of $\mathbb{C}$. It is a strict extension if $k \neq \mathbb{C}$.

Proposition 1 : Let k be a strict topological extension of $\mathbb{C}$ and ν a continuous p-semi-norm on k, $0 < p \leq 1$. Then $\nu = 0$.

We let E be the quotient of k by the kernel of ν, and consider on E the p-norm induced by ν. The quotient mapping $k \to E$ will be called q, and the completion of E will be called $\hat{E}$.

Choose $x \in k \setminus \mathbb{C}$, and $t \in k$. Let $f(z) = t.(x-z)^{-1}$ for $z \in \mathbb{C}$. Then f defined and k-valued on $\mathbb{C}$. This function tends to zero at infinity. We also know that f is of class C_∞ on the complex sphere, and a solution of the Cauchy-Riemann equations.

The function $q \circ f$ is $\hat{E}$-valued, of class C_∞, vanishes at infinity, and is a solution of the Cauchy-Riemann equation. $\hat{E}$ is locally pseudo-convex, $q \circ f$ is a function of class C_∞, with values in $\hat{E}$, when we consider on $\hat{E}$ the small boundedness $\mathcal{B}$ (chapter IV, paragraph 5). This function is therefore analytic, $(\hat{E}, \mathcal{B})$-valued, $\mathcal{B}$ is a convex boundedness, we may apply the Liouville theorem, $q \circ f = 0$.

It is not difficult to show that $f(z) = t.z^{-1} + O(|z|^{-2})$, hence

$q \circ f = q(t)\, z^{-1} + O(|z|^{-2})$, but $q \circ f = 0$ so $q(t) = 0$.
This is the required result.

Corollary 1 : Let k be a strict topological extension of $\mathbb{C}$, and let U be a neighbourhood of the origin in k. Let $0 < p \leq 1$. The absolutely p-convex hull of U is equal to k.

This is true because the p-homogeneous Minkowski functional of $co_p\, U$ is a continuous p-semi-norm, and vanishes identically. If A_n is a sequence of sets and $0 \in A_n$ for all n, we write

$$\Sigma' A_n = \bigcup_n \Sigma_1^n A_k$$

Corollary 2 : Let k be a strict topological extension of $\mathbb{C}$, let U be a neighbourhood of the origin, let d be a positive number. Then

$$k = \Sigma'\, n^{-d}\, U$$

We take U balanced. Let $x \in k$, then $x \in co_p U$ for all values of $p > 0$. This means that x is a finite linear combination

$$x = \Sigma_1^n \lambda_n u_n$$

with $\Sigma|\lambda_n|^p \leq 1$, and for all n : $u_n \in U$. We may assume that $|\lambda_1| \geq \dots \geq |\lambda_n|$, then $|\lambda_k| \geq k^{-1/p}$. We have $x = \Sigma k^{-d} u'_k$ if $u'_k = \lambda_k\, k^{1/p}$ and $1/p = d$.

Note : In a way, corollary 2 is a best possible result. It is possible to find a topological field $(k, \mathcal{T})$ such that a neighbourhood U of the origin can be associated to every decreasing sequence (s_k) with the property

$$\Sigma' s_k\, U \neq k$$

This will be proved at the end of paragraph 4.

The author does not know whether a field topology τ can be associated to such a rapidly decreasing sequence, in such a way that a neighbourhood V of the origin can be associated to every neighbourhood U of the origin, with

$$U \supseteq \Sigma' s_k V$$

2. Fields with discontinuous inverse : It is not difficult to find a strict extension k of $\mathbb{C}$, with a locally convex topology τ , such that multiplication is a continuous mapping $k \times k \to k$. The topology is even complete. Inversion cannot be continuous, because of proposition 1.

k will be the field of germs of meromorphic functions of one variable on neighbourhoods of the origin in $\mathbb{C}$. Then $k = \bigcup E_n$ where E_n is the space of functions f on the disc $|z| \leq n^{-1}$ which are continuous up to the boundary, and such that $z^n f$ is holomorphic on the open disc. We norm E_n by

$$\nu_n(f) = \sup_{|z| \leq 1/n} |z^n f(z)|$$

Then E_n is a Banach space, and the identity mapping $E_n \to E_{n+1}$ is compact.

$k = \bigcup E_n$ is a Silva space. The direct limit topology of k is locally convex and complete. To show that multiplication $k \times k \to k$ is continuous, it is sufficient to show that it is bounded, and this is clear.

A metric topology for which multiplication is continuous can be found on k in the following way. We start out with any continuous semi-norm ν_1 for the topology τ and by induction, assuming that ν_n has been found, we find ν_{n+1} in such a way that

$$\nu_n(x.y) \leq \nu_{n+1}(x)\nu_{n+1}(y)$$

The sequence of semi-norms $\{\nu_n\}$ defined in this way defines a metric topology $\mathcal{T}_1$ on k for which multiplication is continuous.

For no choice of the sequence $\{\nu_n\}$ is the completion of k a field. This follows from Corollary 2 of Proposition 2 in Chapter VII. If the completion were a field inversion would be continuous on its domain and it is not.

3. The Williamson topology on $\mathbb{C}(t)$. Every strict extension of $\mathbb{C}$ contains a subfield isomorphic to $\mathbb{C}(t)$, the field of rational functions of one variable. The first problem in the search for topologies on strict extensions of $\mathbb{C}$ is that of topologizing $\mathbb{C}(t)$. J.H.Willimason [5] described a field topology on $\mathbb{C}(t)$ in the following way.

I is the unit interval $0 \leq x \leq 1$; M(I) is the space of almost everywhere defined measurable functions on I which convergence in measure. We embed $\mathbb{C}(t)$ in M(I) associating to $u(t) = P(t)/Q(t)$ the almost everywhere defined function $u(x) = P(x) / Q(x)$.

With convergence in measure , M(I) is a Fréchet algebra. The set of invertible elements is not open, but it must be a G_δ: it is clear that the mapping $f \to f^{-1}$ is continuous on its domain.
We have proved in this way.

Proposition 2 : The topology induced on $\mathbb{C}$ (t) by convergence in measure is a field topology.

With this topology $\mathbb{C}(t)$ is not complete. Worse, it is not completable. It is clear (because of the Stone Weierstrass theorem) that $\mathbb{C}(t)$ is dense in M(I). The completion is not a field, it does not even have any closed prime ideals. It is not difficult to show that the only closed ideals of M(I) are the ideals α_X of functions which vanish almost everywhere on an almost everywhere defined measurable set X, but α_X is never prime.

This implies that there is no continuous homomorphism of $\mathbb{C}(t)$ with the topology above into a complete topological field k. If there were, the homopmrphism would extend to the completion, and the kernel would be a closed prime ideal of M(I).

We could try to get better results, embedding $\mathbb{C}(t) \subseteq M(I_2)$ where $I_2 = I + iI$, associating again to $u = P/Q \in \mathbb{C}(t)$ the mapping $x \to P(x)/Q(x)$, now defined almost everywhere on $\mathbb{C}$. But $\mathbb{C}(t)$ is just as dense in $M(I_2)$ as in M(I). If j is a closed, simple, rectifiable curve, and if (j) is its interior

$$1_{(j)}(x) = \frac{1}{2\pi i} \int_j \frac{ds}{s - x}$$

The integral converges in measure, so $1_{(j)}$, the characteristic function of the interior of j, is in the closure of $\mathbb{C}(t)$. Step functions are dense in $M(I_2)$, so $\mathbb{C}(t)$ is dense.

It is well known that M(I) and $M(I_2)$ are isomorphic topological algebras, so that these two topologies on $\mathbb{C}(t)$ share their undesirable properties as much as their desirable ones.

<u>4. The strongest topology of $\mathbb{C}(t)$.</u> : It is possible to find a Hausdorff field topology on $\mathbb{C}(t)$ which induces on $\mathbb{C}$ the usual topology. The initial topology of all continuous homomorphisms $u : \mathbb{C}(t) \to k$, where k is a topological field , and u is continuous on $\mathbb{C}$, is the strongest field topology $\mathcal{T}$ on $\mathbb{C}(t)$ which induces on $\mathbb{C}$ the given topology.

Let $D = \{z \mid |z| \leq 1\}$ be the closed unit disc in $\mathbb{C}$. Then $D \cup \{t\}$ is compact, metrizable, and generates $\mathbb{C}(t)$. Also, $\mathcal{T}$ is the strongest field topology on $\mathbb{C}(t)$ which induces on $D \cup \{t\}$ the given topology.

Because $D \cup \{t\}$ is metrizable, it generates a countable compactology on $\mathbb{C}(t)$ (Chapter 3, proposition 10), and $\mathcal{T}$ is the direct limit topology for this compactology. We know that $\mathcal{T}$ is sequentially complete (Chapter 3, proposition 7).

$\mathcal{T}$ is not metrizable. If it were, $\mathbb{C}(t)$ would be complete since it is sequentially complete. One of the elements of the countable compactology we are considering would have an interior. $\mathbb{C}(t)$ would be locally compact. But $\mathbb{C}(t)$ is an infinite dimensional topological vector space.

The compactology of $\mathbb{C}(t)$ is not too difficult to describe. Consider any submultiplicative norm on the ring $\mathbb{C}[t]$ of polynomials in t, e.g. $||\Sigma\, a_k t^k|| = \Sigma |a_k|$, let $d^\circ P$ denote the degree of the polynomial P, and write

$$B_n = \{\frac{P(t)}{Q(t)} \mid d^\circ P \leq n,\ d^\circ Q \leq n,\ ||P|| \leq n,\ ||Q|| = 1\}$$

It is obvious that B_n belongs to the field compactology generated by $D \cup \{t\}$.

We shall show that these sets are the basis of a field compactology. If $P/Q \in B_n, P'/Q' \in B_n$, we have

$$\frac{P}{Q} + \frac{P'}{Q'} = \frac{PQ' + P'Q}{QQ'}$$

$$\frac{P}{Q} \cdot \frac{P'}{Q'} = \frac{PP'}{QQ'}$$

Using the fact that $\|Q.Q'\| \geqslant \varepsilon_n$ if $\|Q\| = \|Q'\| = 1$, $d^\circ Q \leqslant n$, $d^\circ Q' \leqslant n$, we find a scalar M_n such that $B_n + B_n \subseteq M_n B_{2n}$, $B_n . B_n \subseteq M_n B_{2n}$. This already shows that we have a ring compactology on $\mathbb{C}(t)$.

A closed subset of B_n which does not contain zero is contained in

$$B_{n.\varepsilon} = \{\frac{P(t)}{Q(t)} \Big| \; d^\circ P \leqslant n, \; d^\circ Q \leqslant n, \; \varepsilon \leqslant \|P\| \leqslant n, \; \|Q\| = 1\}$$

for some $\varepsilon > 0$, and $B_{n.\varepsilon}{}^{-1} \subseteq (1/n\varepsilon) B_n$, so that we do have a field compactology on $\mathbb{C}(t)$.

The compactology we have found on $\mathbb{C}(t)$ is non convex but of Silva type. This follows from the relation

$$\frac{P'}{Q'} - \frac{P}{Q} = \frac{P'Q - PQ'}{QQ'}$$

Assume that $P_k/Q_k \to P/Q$ with $P_K/Q_k \in B_n$. We may assume that $P_k \to P$, $Q_k \to Q$, $\|P_k\| \leqslant n$, $\|Q_k\| = 1$, $d^\circ P_k \leqslant n$, $d^\circ Q_k \leqslant n$.
Then $P_k/Q_k - P/Q \in \varepsilon_k B_{2n}$ for some sequence of $\varepsilon_k \to 0$.

Among other things, this implies that the topology $\mathcal{T}$ is the strongest vector space topology on $\mathbb{C}(t)$ for which the sets B_n are all bounded.

We can introduce smaller sets that the sets B_n. Let

$$C = \{t\} \cup \{1\} \cup \{\frac{1 + |z|}{t - z}\} \Big| \; z \in C\}$$

Let

$$C^n = \{f_1 \dots f_n \; \Big| \; \forall \, i : f_i \in C\}$$

Then B_n is contained in a finite linear combination with bounded coefficients of elements of C^{2n}, so that it is sufficient that the sets C^n be bounded.

C'^n will be balanced hull of C^n. The topology τ is the strongest vector space topology on $\mathbb{C}(t)$ for which the sets C'^n all are bounded. We obtain a fundamental system of neighbourhoods of the origin in $\mathbb{C}(t)$ by associating to every sequence of strictly positive numbers $\{\varepsilon_n\}$ the set

$$V(\{\varepsilon_n\}) = \Sigma' \varepsilon_n C'^n$$

A linear mapping $\mathbb{C}(t) \to E$, with for E a topological vector space, is continuous if it maps each C^n onto a bounded subset of E.

Note. At the end of paragraph 1, we mentioned the fact that a result we have proved was "best possible". Proof of this uses the ring

$$\mathfrak{a} = \cup_{p>0} L_p(I)$$

with its direct limit topology. It is not difficult to find, for each rapidly decreasing sequence (s_k) and each neighbourhood U of the origin a neighbourhood V of the origin in $\mathfrak{a}$ such that

$$U \supseteq \Sigma'_n \varepsilon_n V$$

It is also clear that $\mathbb{C}(t) \subseteq \cup L_p(I)$ and that each set C^n is bounded in $L^p(I)$ for some p. The topology τ of $\mathbb{C}(t)$ is therefore stronger than the topology induced by $\cup L_p$ on $\mathbb{C}(t)$. This is the required example.

5. A complete metric extension : We want to show now that $\mathbb{C}$ has a complete, metrizable, strict extension. The existence of such an

extension was mentioned as an open problem by W.Żelasko [99], paragraph 7. The construction is long and technical.

a) $\mathbb{C}^* = \mathbb{C} \cup \{\infty\}$ will be the complex sphere. We map $\mathbb{C}^*$ onto a sphere $S_2 \subseteq \mathbb{R}^3$ by a stereographic projection, and we let $d(x, y)$ be the spherical distance of two elements of $\mathbb{C}^*$, or better the arcwise distance of their images in S_2. All metric statements involving $\mathbb{C}^*$ will refer to the distance d.

Let now $E \subseteq \mathbb{C}^*$, and $0 < p \leq 1$. We define $\gamma_p(E)$ putting $\gamma_p(E) < a$ iff $E \subseteq \bigcup_1^\infty D_i$, each D_i a disc of radius r_i and $\Sigma_1^\infty r_i^p < a$. The set function γ_p is countably subadditive, obviously

$$\gamma_p(\bigcup_1^\infty E_n) \leq \Sigma_1^\infty \gamma_p(E_n)$$

This set function is not unrelated to Hausdorff measure and dimension.

For example, the set E will have Hausdorff dimension zero iff $\gamma_p E = 0$ for all p. Such sets E will be called <u>negligible</u>.

$\mathfrak{A}$ will be a Banach algebra with a single maximal ideal. When $a \in \mathfrak{A}$, we shall $\hat{a}$ be the value in a of the unique character of $\mathfrak{A}$, so $\hat{a} \in \mathbb{C}$. A mapping $u : \mathbb{C}^* \to \mathfrak{A}$ will be called negligible if it vanishes everywhere except maybe on a negligible set. $M_\gamma(\mathbb{C}^*, \mathfrak{A})$ will be the quotient by the ideal of negligible functions, of the algebra of $u : \mathbb{C}^* \to \mathfrak{A}$ such that, for all p and ε, one can find a set E with $\gamma_p(E) < \varepsilon$ the function u being continuous on the complement of E.

The following neighbourhoods of the origin will topologize $M_\gamma(\mathbb{C}^*, \mathfrak{A}) : U(\varepsilon, \eta, p) = \{f \mid \gamma_p(\{\|f\|) \geq \varepsilon\}) < \eta\}$

Then $M_\gamma(\mathbb{C}^*, \mathcal{A})$ is a complete metrizable algebra. Further, Egoroff's theorem shows that it is possible to extract from each convergent sequence f_n a subsequence which converges "almost uniformly", uniformly on the complements of sets E_k with $\gamma_p(E_k) \to 0$ for all p.

Inversion is continuous on its domain in $M_\gamma(\mathbb{C}^*, \mathcal{A})$ because $f^{-1} - 1 \in U(\varepsilon/(1 - \varepsilon), \eta, p)$ when f has an inverse and $f - 1 \in U(\varepsilon, \eta, p)$. Also, f has an inverse in $M_\gamma(\mathbb{C}^*, \mathcal{A})$ when the set of $s \in \mathbb{C}^*$ such that $f(s)$ is not invertible is a negligible set. Since $\mathcal{A}$ has a single maximal ideal, f has an inverse iff $\hat{f}$ does not vanish except on a negligible set.

We choose next a sequence $n \to c_n$ of positive real numbers, in such a way that $c_o = 1$, c_{n+1}/c_n is a decreasing sequence tending to zero, and

$$\Sigma \frac{1}{n} \frac{c_{n+1}}{c_n} = \infty$$

These properties will hold, for example, if

$$\frac{c_{n+1}}{c_n} = \frac{1}{\log(n + 2)}$$

$\mathcal{A}$ will be the set of complex sequences $\{a_n\}$ such that

$$||a|| = \Sigma\, c_n\, |a_n| < \infty$$

This is a radical Banach algebra for convolution.

If f is a function of class C_∞ on $\mathbb{R}$, if $a_n(s) = f^{(n)}(s)/n!$ is the nth Taylor coefficient of f, and $a(s) = (a_n(s))$ is the sequence of Taylor coeeficients of a, then the statement that $s \to a(s)$

is a bounded $\mathfrak{a}$-valued function implies that f is in the "class of functions" determined by the constants $n!\, c_n^{-1}$.
This class is quasi-analytic because $\Sigma\, c_{n+1}/n\, c_n$ diverges. If f vanishes with all its derivatives at one point of $\mathbb{R}$, then f vanishes identically.

Let now $r \in \mathbb{C}(t)$, $z \in \mathbb{C}$, z not a pole of r. We let $\tilde{r}(z)$ be the sequence of Taylor coefficients of r at the point z, i.e.

$$\tilde{r}_n(z) = \frac{r^{(n)}(z)}{n!}$$

Then $\tilde{r}$ is defined on the complement of a finite set in $\mathbb{C}^*$, the mapping $z \to \tilde{r}(z)$ is a continuous $\mathfrak{a}$-valued function on its domain.
The mapping $r \to \tilde{r}$ is an imbedding of $\mathbb{C}(t)$ in $M_\gamma(\mathbb{C}^*, \mathfrak{a})$. We shall identify r and $\tilde{r}$.

We want to show that the closure $\overline{\mathbb{C}(t)}$ of $\mathbb{C}(t)$ in $M_\gamma(\mathbb{C}^*, \mathfrak{a})$ is a topological field. We already know that the algebraic operations are continuous, including inversion on its domain. Also, an element of $\overline{\mathbb{C}(t)}$ has an inverse in $\overline{\mathbb{C}(t)}$ if it has one in $M_\gamma(\mathbb{C}^*, \mathfrak{a})$. Let $r_n \in \mathbb{C}(t)$, $r_n \to f$, then $1/r_n \in \mathbb{C}(t)$, $1/r_n \to 1/f$ because the inverse is continuous on its domain in M_γ. This shows that $1/f \in \overline{\mathbb{C}(t)}$.

b) We will therefore have an example of a complete metric field if we can show that the vanishing set of $\hat{f}$ is negligible, has zero Hausdorff dimension, when f is in the closure of $\mathbb{C}(t)$ in $M_\gamma(\mathbb{C}^*, \mathfrak{a})$. The proof will use the theory of quasi-analytic functions. We shall have to find great circles which lie within the domain of f.

It makes sense to speak of great circles of $\mathbb{C}^*$ since $\mathbb{C}^*$ has been identified with a sphere $S_2 \subseteq \mathbb{R}^3$ by means of a stereographic projection.

Let $F \in \overline{\mathbb{C}(t)}$. Choose $p > 0$, $\eta > 0$. It is possible to find a sequence $r_1, \ldots, r_n, \ldots$ of elements of $\mathbb{C}(t)$ which tends to f uniformly on the complement of a set E with $\gamma_p(E) < \eta$.

The set E is contained in the union of a sequence of discs D_i, of radius ρ_i and $\Sigma \rho_i^p < \eta$.

We shall let D_i' be the union of two antipodal discs, of radius $\rho_i^{1/2}$ each, one of these discs concentric with D_i. We define next E' as the union of the D_i'. Clearly $\gamma_{2p}(E') < 2\eta$.

Let $z \notin E'$, let z' be its antipode. To show that there exist great circles through z and z' which do not meet E, we shall estimate the angular measure of the set of great circles through z and z' which meet E and show that this is less than π.

Let α_i be the center of the circle D_i, let a_i be the smaller of $d(z, a_i)$, $\pi - d(z, a_i)$. Then $a_i > \rho_i^{1/2}$ since otherwise z would belong to D'. The law of sines in spherical geometry shows that the angular measure of the set of great circles which meet D_i is

$$2 \text{ arc sin } \frac{\sin \rho_i}{\sin a_i}$$

Applying the inequalities arc sin $t \leqslant \pi t/2$, $\sin \rho_i \leqslant \rho_i$, $\sin a_i \geqslant 2a_i/\pi$ (because $a_i < \pi/2$) and $a_i \geqslant \rho_i^{1/2}/2$, and the angular measure of the great circles which meet E is at most

$$\frac{\pi^2}{2} \Sigma \rho_i^{1/2} = \theta$$

Choosing p and η small enough, we can make this as small as we wish.

We must also estimate the geodesic distance between z_1 and z_2 in $\mathbb{C}^* \setminus E$, i.e. the infimum of the lengths of rectifiable curves connecting z_1 and z_2 through $\mathbb{C}^*\setminus E$, when z_1 and z_2 both belong to $\mathbb{C}^*\setminus E'$. It will be sufficient to have a local result, we assume that $d(z_1, z_2)<1$.

Consider the arc of great circle through z_1 and z_2. It is possible to find a great circle through z_1, on the left of the first great circle, which does not meet E, at an angle at most θ of the great circle arc (z_1, z_2). It is also possible to find a great circle through z_2, which does not meet E, and makes an angle at most θ with the great circle (z_2, z_1), this second great circle being on the right of the great circle (z_2, z_1).

Let ζ be the intersection of these two great circles, let $\delta_1 = d(z_1, \zeta)$, $\delta_2 = d(z_2, \zeta)$. Then $\delta_1 + \delta_2$ is the length of a path which does not meet E and connects z_1 and z_2. Also

$$\delta_1 + \delta_2 < \arcsin\left(\frac{\sin d}{\sin \theta}\right) \leqslant A\, d$$

where A is a positive constant (A depends on θ, but θ has been chosen).

The proof of our estimate for $\delta_1 + \delta_2$ involves some spherical trigonometry, and a convexity argument on the arc sine function. We shall not give it. The only result we need is the estimate $\delta_1 + \delta_2 < A\, d$ for some finite A, and that should be clear.

c) We now return to the function $f \in M_\gamma(\mathbb{C}^*, \mathfrak{a})$ that we were considering. On $C^*\setminus E$, it is a limit of rational functions. We shall show that f can only have a finite number of zeroes on $\mathbb{C}^*\setminus E'$, unless it vanishes identically. Of course, when we speak of zeroes of f, we think of zeroes of $\hat{f}$.

In each $s \in \mathbb{C}^*\setminus E$, f is a sequence $(f_0(s), f_1(s), \ldots)$. The point s is a zero of order k if $f_0(s) = \ldots f_{k-1}(s) = 0$, while $f_k(s) \neq 0$. It is a zero of infinite order if all $f_k(s)$ vanish. We want to show that the zeroes of finite order are isolated in $\mathbb{C}^*\setminus E'$. There can only be a finite number of such points. And that f cannot have any zero of infinite order in E' unless it vanishes identically.

We shall start out with the case where f has a zero of infinite order in $D\setminus E'$. Let z_1 be such a point, and j be a great circle through z_1 which does not meet E. On j, f belongs to a quasi-analytic class, so f vanishes identically along with all its derivatives. Let then z_2 be another point in $D\setminus E'$, let j' be a great circle through z_2 which does not meet E. The restriction of f to j' also belongs to a quasi-analytic class, and vanishes along with all its derivatives at the intersection of j and j', so f vanishes identically on j' and vanishes therefore at z_2.

Consider now a point $s \in \mathbb{C}^*\setminus E$, $s \neq \infty$, where f has a zero of finite order, k. Let $z \in \mathbb{C}^*\setminus E'$ be in the neighbourhood of s, and j be a rectifiable path from s to z of length less than $A\, d(s,z)$. Let r be a rational function without a pole on the path j. Then

$$r(z) = \Sigma_0^k\, r_k(s)(z - s)^k + \int_j r_{k+1}(t)(z - t)^k\, dt$$

But f is a uniform limit on j of such functions, the f_k are limits of the r_k too. At the limit $f_0(s) = \ldots = f_{k-1}(s) = 0$,

$f(z) = f_k(s)(z - s)^k + \int_j r_{k+1}(t)(z - t)^k\, dt$

$$= f_k(s)(z - s)^k + o(|z - s|^k)$$

This shows that $f(z) \neq 0$ if $z \neq s$ is near enough to s.

This concludes the proof, $\overline{\mathbb{C}(t)}$ is a complete metric field.

Notes and remarks : The subject of this chapter has already been surveyed by R.Arens [6], J.H.Williamson [94] and W.Żelazko [99], paragraph 8. We tried to include the significant results in these surveys and some more recent results.
I.M.Gelfand [29] and S.Mazur [52] showed that a strict extension of $\mathbb{C}$ could not have an algebra norm. R.Arens [6] showed that the continuous linear forms on such an extension vanish identically.
W.Żelazko [95] generalizes the Gelfand-Mazur theorem, showing that such a strict extension did not have an algebra p-norm. Proposition 1 is due to P.Turpin and the author [76]. It is a simultaneous generalization of the Arens and the Żelazko theorems.

To show that the Arens theorem is effectively a generalization of the Gelfand-Mazur theorem, and proposition 1 of Żelazko's, we must remember that the mapping $a \to a^{-1}$ is continuous on its domain when the algebra is normed or p-normed. Continuity of the inverse does not therefore have to be postulated.

The results of paragraph 2 and 3, at least the existence proofs of topologies on $\mathbb{C}(t)$ having desirable properties, are due to Williason [94]. Williamson also gives an explicit sequence of semi-norms on $\mathbb{C}(t)$ for which this is a metrizable locally convex algebra.

To the author's best knowledge, the results of paragraphs 4 and 5 are new, and due to him. Żelazko ([99], paragraph 7) mentions as open the problem of whether a complete metric division algebra exists.

BIBLIOGRAPHY

1. N.ADASCH. Topologische Produkte gewisser topologische Vektorräume. To be published Math.Ann.

2. N.ADASCH. Der Graphensatz in topologischen Vektorräume. To be published Math.Zeitschrift.

3. N.ADASCH. Vollständigkeit und der Graphensatz. To be published.

4. G.ALLAN. A spectral theory for locally convex algebras. Proc. London Math. Soc. (3)15.1965. p.399-421.

5. T.AOKI. Locally bounded topological vector spaces. Proc.Imp. Acad. Tokyo. 18.1942. p.599-594.

6. R.ARENS. Linear topological division algebras. Bull.Am.Math.Soc. 53.1947. p.623-630.

7. R.ARENS. The group of invertible elements of a commutative Banach algebra. Studia Math. Seria Spec.Z.I.1963. p.21-23.

8. R.ARENS. To what extent does the space of maximal ideals determine the algebra ? in Function algebras. Scott Foresman and Co.1966. p.165-168.

9. R.ARENS and A.P.CALDERON. The analytic functional calculus of several Banach algebra elements. Ann.of Math.62.1955. p.204-216.

10. N.BOURBAKI.Espaces vectoriels topologiques. Paris.Hermann et Cie. 1953, 1955.

11. N.BOURBAKI.Théories spectrales. Chapitre I. Algèbres normées. Paris. Hermann et Cie.1967.

12. H.BUCHWALTER. Espaces vectoriels bornologiques. Pub.Dép.Math.Lyon. 2.1965.p.1-53.

13. H.BUCHWALTER. Topologies, bornologies et compactologies. Thèse. Dép.Math.Lyon.1968.

14. H.CARTAN. Idéaux et modules de fonctions analytiques de plusieurs variables complexes. Bull.Soc.Math.France.78.1950.p.29-64.

15. I.CNOP. A theorem concerning holomorphic function with bounded growth. Thesis. Vrije Universiteit Brussel.1971.

16. F.CNOP-GRANDSARD. Sur les fonctions de type ($\mathcal{S}$) à valeurs dans un b-espace. Bull.Acad.Royale Belgique.Cl.des Sc.56.1970. p.138-143.

17. H.COOK and H.R.FISCHER. On equicontinuity and continuous convergence. Math.Ann.159.1965.p.94-104.

18. M.DE WILDE. Résaux dans les espaces linéaires à semi-normes. Mémoires Société Royale des Sciences de Liège.1969.

19. D.O.ETTER Jr. Vector-valued analytic functions.Trans.Am.Math.Soc. 119.1965.p.352-366.

20. A.V.FERREIRA. Some remarks on b-spaces. Portugaliae Math.26.1967. p.421-447.

21. J.P.FERRIER. Ensembles spectraux et approximation polynomiale pondérée. Bull.Soc.Math.France.96.1968.p.289-365.

22. J.P.FERRIER. Sur la convexité holomorphe et les limites inductives d'algèbres $\mathcal{O}(\delta)$.C.R.Acad.Sci.Paris.

23. J.P.FERRIER. Approximation with bounds of holomorphic functions of several complex variables. (to be published).

24. H.R.FISCHER. Limesräume. Math.Ann.137.1959.p.269-303.

25. G.MARINESCU. Espaces polynormés, duals des espaces localement convexes.C.R.Acad.Sci.Paris.241.1955.p.1693-1695.

26. C.FOIAS and G.MARINESCU. Sur le prolongement des fonctionnelles linéaires dans les espaces vectoriels pseudo-topologiques. C.R.Acad.Sci.Paris.254.1962.p.2274-2276.

27. C.FOIAS and G.MARINESCU. Fonctionnelles linéaires dans les réunions dénombrables d'espaces de Banach réflexifs.C.R.Acad. Sci.Paris.261.1965.p.4958-4960.

28. A.FRÖLICHER and W.BUCHER. Calculus in vector spaces without norm. Springer lectures notes in Mathematics.30.1960.

29. I.M.GELFAND. Normierte Ringe.Math.Sbornik.9(51) 1941.p.3-24.

30. G.GLAESER. Etude de quelques algèbres tayloriennes. J. Analyse Math. 6; 1958. p. 1-124.

31. B. GRAMSCH. Integration und holomorphe Funktionen in lokalbeschränkten Räumen. Math Ann. 162. 1965. p. 190-210.

32. B. GRAMSCH. Tensorprodukte und Integration vektorwertigen Funktionen. Math. Zeitschrift. 100. 1967. p. 106-122.

33. B. GRAMSCH. Funktionalkalkül mehrerer Veränderlichen in lokalbeschränkten Algebren. Math. Ann. 174. 1967. p. 311-344.

34. B. GRAMSCH. Integration in topologischen Vektorräume und lokal p-konvexen Algebren. in General topology and its relation to Analysis and algebra ; proceedings of the second Prague symposium. 1968. p. 147-155.

35. H.GRAUERT. Holomorphe Funktionen mit Werten in komplexen Lieschen Gruppen.Math.Ann. 133.1957.p.450-472.

36. H.GRAUERT. Analytische Faserungen über holomorph-vollständiger Räumen.Math.Ann. 135.1958.p.263-273.

37. A.GROTHENDIECK. Produits tensoriels topologiques et espaces nucléaires. Mémoires Am.Math.Soc. 1954.

38. A.GROTHENDIECK. Espaces vectoriels topologiques. São Paulo.1958.

39. R.GUNNING and H.ROSSI. Analytic functions of several complex variables. Prentice Hall.1965.

40. H.HOGBE-NLEND. Complétion, tenseurs, et nucléarité en bornologie. In publication, J.de Math.P. et App. (see also Thèse, Bordeaux, 1969).

41. H.HOGBE-NLEND. Les fondements de la bornologie. In Publication, Springer Lecture Notes in Mathematics.

42. L.HÖRMANDER. An introduction to complex analysis. Van Nostrand. 1966.

43. S.O.IYAHEN. On certain classes of linear topological spaces. Proc. London Math.Soc. (3)18.1968.p.285-307.

44. V.KLEE. Shrinkable neighbourhoods in Hausdorff linear spaces. Math.Ann.141.1960.p.281-285.

45. V.KLEE. Leray-Schauder theory without local convexity. Math.Ann. 141-1960.p.208-296.

46. J.KÖHN. Induktive Limiten nichlokalkovexer topologischen linearen Räumen. Math.Ann.181.1969.p.269-278.

47. T.KŌMURA and Y.KŌMURA. Über die Einbettung der nuklearen Räume in $(s)^A$. Math.Ann.162.1965-66.p.284-288.

48. G.KÖTHE. Die Teilräume eines linearen Koordinatenraumes. Math.Ann. 114.1937.p.99-125.

49. G.KÖTHE. Topologische lineare Räume. Springer Verlag.1960.

50. J.LERAY. Fonction de variables complexes : sa représentation comme somme de puissances négatives de fonctions linéaires.R.C.Acad. Lincei. série 8.v.20.1956.p.589-590.

51. E.R.LORCH. The theory of analytic functions in normed abelian vector rings.Trans.Am.Math.Soc.53.1942.p.238-248.

52. S.MAZUR. Sur les anneaux linéaires.C.R.Acad.Sci.Paris.207.1938. p.1025-1027.

53. S.MAZUR and W.ORLICZ. Sur les espaces métriques linéaires.I. Studia Math.10.1948.p.184-208.

54. S.MAZUR and W.ORLICZ. Sur les espaces métriques linéaires.II. Studia Math.13.1953.p.137-179.

55. E.A.MICHAEL. Locally multiplicatively convex division algebras. Mémoirs Am.Math.Soc.11.1952.

56. J.MIKUSINSKI. Operational calculus. London.Pergamon Press.1959.

57. J.MIKUSINSKI. Distributions à valeurs dans les réunions d'espaces de Banach. Studia Math.19.1960.p.251-285.

58. B.MITJAGIN, S.ROLEWICZ, and W.ŻELASKO. Entire functions in B_o-algebras. Studia Math.21.1962.p.291-306.

59. L.D.NEL. Note on completeness in a pseudo-topological linear space. J.London. Math.Soc.40.1965.p.497-498.

60. G. NOËL. Une immersion de la catégorie des espaces bornologiques convexexes séparés dans une catégorie abélienne. C.R. Acad. Sci. Paris. 269. 1969. p. 238-240.

61. G. NOËL. Produit tensoriel et platitude des Q-espaces. Bulletin Soc. Math. Belgique. 22. 1970. p. 119-142.

62. K. OKA. Sur les fonctions analytiques de plusieurs variables. VII. Sur quelques notions arithmétiques. Bull. Soc. Math. France. 78. 1950. p. 147-186.

63. C. HOUZEL. (editor) Seminaire Banach. (Mimeographed). Ecole Normale Supérieure. Paris. 1963.

64. D. PRZEWORSKA-ROLEWICZ and S. ROLEWICZ. On Integrals of functions with values in a complete linear metric space. Studia Math. 26. 1966. p. 121-131.

65. W. ROBERTSON. Completions of topological vector spaces. Proc. London Math. Soc. 8. 1958. p. 242-257.

66. S. ROLEWICZ. On a certain class of linear metric spaces. Bull. Acad. Pol. Sci. III. 5. 1957. p. 471-473.

67. L. SCHWARTZ. Théorie des distributions. Paris. Hermann et Cie. 1951.

68. J. SEBASTIAO E SILVA. Su certe classi di spazi localmente convessi importanti per li applicazioni. R.C. Math. P. e App. 14. 1955. p. 388-410.

69. J. SEBASTIAO E SILVA. Sur le calcul symbolique d'opérateurs permutables à spectre vide ou non borné. Ann. di Math. Pura e App. 58. 1962. p. 219-276.

70. A.H. SHUCHAT. Approximation of vector-valued continuous functions. (to be published).

71. G.E. SILOV. On the decomposition of a commutative normed ring in a direct sum of ideals. (Russian) Math. Sbornik. 32(74) 1953. p. 353-364.

72. W.SŁOWIKOWSKI. Fonctionnelles linéaires dans les réunions dénombrables d'espaces de Banach réflexifs. C.R. Acad. Paris. 262. 1966. p. 870-872.

73. P. TURPIN. Une remarque sur les algèbres à inverse continu. C.R. Acad. Sci. Paris. 270. 1970. p. 1686-1689.

74. P. TURPIN and L. WAELBROECK. Sur l'approximation des fonctions différentiables à valeurs dans les espaces vectoriels topologiques. C.R. Acad. Sci. Paris. 267. 1968. p. 94-97.

75. P.TURPIN and L. WAELBROECK. Intégration et fonctions holomorphes dans les espaces localement pseudo-convexes. C.R. Acad. Scie. Paris. 267. 1968. p. 160-162.

76.P. TURPIN and L. WAELBROECK. Algèbres localement pseudo-convexes à inverse continu. C.R. Acad. Sci. Paris. 267- 1968. p. 194-195.

77. C.DE LA VALLEE-POUSSIN. Intégrales de Lebesgue, fonctions d'ensembles, classes de Baire.Paris.Gauthiers-Villars et Cie.1916.

78. D.VOGT. Integrations theorie in p-normierten Räumen.Math.Ann.173. 1967.p.219-232.

79. L.WAELBROECK. Le calcul symbolique dans les algèbres commutatives. J. de Math.P.et App.33.1954.p.147-186.

80. L.WAELBROECK. Algèbres commutatives : éléments réguliers. Bull. Soc.Math.Belgique.9.1957.p.42-49.

81. L.WAELBROECK. Etude spectrale des algèbres complètes. Acad.Royale Belgique.Mém.Cl.des Sci.1960.

82. L.WAELBROECK. Les espaces à bornés complets. Colloque sur l'Analyse Fonctionnelle.C.B.R.M. Louvain.1961.p.51-55.

83. L.WAELBROECK. Le complété et le dual d'un espace "à bornés".C.R. Acad.Sci.Paris.253.1961.p.2827-2828.

84. L.WAELBROECK. Continuous inverse locally pseudo-convex algebras. in Summer School on Topological Algebra Theory (Mimeographed) Presses Universitaires de Bruxelles.1967.p.128-185.

85. L.WAELBROECK. Some theorems about bounded structures.J.of Funct. Analysis.1.1967.p.392-408.

86. L.WAELBROECK. Differentiable mappings into b-spaces. J.of Funct. Analysis.1.1967.p.409-418.

87. L.WAELBROECK. Fonctions differentiables et petite bornologie. C.R.Acad.Sci.Paris.267.1968.p.220-222.

88. L.WAELBROECK. Sur les compactologies dénombrables. Séminaire d'Analyse Fonctionnelle de Bordeaux.1969-70. (Mimeographed).

89. L.WAELBROECK. Un espace compactologique non séparé topologiquement. Séminaire d'Analyse Fonctionnelle de Bordeaux.1969-70.(Mimeographed).

90. A.WEIL. L'intégrale de Cauchy et les fonctions de plusieurs variables. Math.Ann.111.1935.p.178-182.

91. H.WHITNEY. Analytic extensions of differentiable functions defined on closed sets. Trans.Am.Math.Soc.36.1934.p.63-89.

92. H.WHITNEY. On the extension of differentiable functions. Bull. Am.Math.Soc.50.1944.p.76-81.

93. A.WILANSKI. Topics in functional analysis. Springer Lecture Notes in Mathematics.45.1967.

94. J.H.WILLIAMSON. On topologising the field $\mathbb{C}(t)$. Proc.Am.Math.Soc. 5.1954.p.729-734.

95. W.ŻELAZKO. On the locally bounded and m-convex topological algebras. Studia Math.19.1960.p.333-356.

96. W.ŻELAZKO. On the radicals of p-normed algebras. Studia Math.21. 1962.p.203-206.

97. W.ŻELAZKO. Analytic functions in p-normed algebras. Studia Math. 21.1962.p.345-350.

98. W.ŻELAZKO. On the decomposition of a commutative p-normed algebra into a direct sum of ideals. Coll.Math.10.1953.p.57-60.

99. W.ŻELAZKO. Metric generalizations of Banach algebras. Rozprawy Math.Warsaw.47.1965.

100. Les espaces nucléaires. Quatrième Foire Estivale d'Analyse Fonctionnelle. 1969. Mathematics Department, University of Brussels.

Vol. 111: K. H. Mayer, Relationen zwischen charakteristischen Zahlen. III, 99 Seiten. 1969. DM 8,–

Vol. 112: Colloquium on Methods of Optimization. Edited by N. N. Moiseev. IV, 293 pages. 1970. DM 18,–

Vol. 113: R. Wille, Kongruenzklassengeometrien. III, 99 Seiten. 1970. DM 8,–

Vol. 114: H. Jacquet and R. P. Langlands, Automorphic Forms on GL (2). VII, 548 pages. 1970. DM 24,–

Vol. 115: K. H. Roggenkamp and V. Huber-Dyson, Lattices over Orders I. XIX, 290 pages. 1970. DM 18,–

Vol. 116: Sèminaire Pierre Lelong (Analyse) Année 1969. IV, 195 pages. 1970. DM 14,–

Vol. 117: Y. Meyer, Nombres de Pisot, Nombres de Salem et Analyse Harmonique. 63 pages. 1970. DM 6.–

Vol. 118: Proceedings of the 15th Scandinavian Congress, Oslo 1968. Edited by K. E. Aubert and W. Ljunggren. IV, 162 pages. 1970. DM 12,–

Vol. 119: M. Raynaud, Faisceaux amples sur les schémas en groupes et les espaces homogènes. III, 219 pages. 1970. DM 14,–

Vol. 120: D. Siefkes, Büchi's Monadic Second Order Successor Arithmetic. XII, 130 Seiten. 1970. DM 12,–

Vol. 121: H. S. Bear, Lectures on Gleason Parts. III, 47 pages. 1970. DM 6,–

Vol. 122: H. Zieschang, E. Vogt und H.-D. Coldewey, Flächen und ebene diskontinuierliche Gruppen. VIII, 203 Seiten. 1970. DM 16,–

Vol. 123: A. V. Jategaonkar, Left Principal Ideal Rings. VI, 145 pages. 1970. DM 12,–

Vol. 124: Séminare de Probabilités IV. Edited by P. A. Meyer. IV, 282 pages. 1970. DM 20,–

Vol. 125: Symposium on Automatic Demonstration. V, 310 pages. 1970. DM 20,–

Vol. 126: P. Schapira, Théorie des Hyperfonctions. XI, 157 pages. 1970. DM 14,–

Vol. 127: I. Stewart, Lie Algebras. IV, 97 pages. 1970. DM 10,–

Vol. 128: M. Takesaki, Tomita's Theory of Modular Hilbert Algebras and its Applications. II, 123 pages. 1970. DM 10,–

Vol. 129: K. H. Hofmann, The Duality of Compact Semigroups and C*- Bigebras. XII, 142 pages. 1970. DM 14,–

Vol. 130: F. Lorenz, Quadratische Formen über Körpern. II, 77 Seiten. 1970. DM 8,–

Vol. 131: A Borel et al., Seminar on Algebraic Groups and Related Finite Groups. VII, 321 pages. 1970. DM 22,–

Vol. 132: Symposium on Optimization. III, 348 pages. 1970. DM 22,–

Vol. 133: F. Topsøe, Topology and Measure. XIV, 79 pages. 1970. DM 8,–

Vol. 134: L. Smith, Lectures on the Eilenberg-Moore Spectral Sequence. VII, 142 pages. 1970. DM 14,–

Vol. 135: W. Stoll, Value Distribution of Holomorphic Maps into Compact Complex Manifolds. II, 267 pages. 1970. DM 18,–

Vol. 136: M. Karoubi et al., Séminaire Heidelberg-Saarbrücken-Strasbuorg sur la K-Théorie. IV, 264 pages. 1970. DM 18,–

Vol. 137: Reports of the Midwest Category Seminar IV. Edited by S. MacLane. III, 139 pages. 1970. DM 12,–

Vol. 138: D. Foata et M. Schützenberger, Théorie Géométrique des Polynômes Eulériens. V, 94 pages. 1970. DM 10,–

Vol. 139: A. Badrikian, Séminaire sur les Fonctions Aléatoires Linéaires et les Mesures Cylindriques. VII, 221 pages. 1970. DM 18,–

Vol. 140: Lectures in Modern Analysis and Applications II. Edited by C. T. Taam. VI, 119 pages. 1970. DM 10,–

Vol. 141: G. Jameson, Ordered Linear Spaces. XV, 194 pages. 1970. DM 16,–

Vol. 142: K. W. Roggenkamp, Lattices over Orders II. V, 388 pages. 1970. DM 22,–

Vol. 143: K. W. Gruenberg, Cohomological Topics in Group Theory. XIV, 275 pages. 1970. DM 20,–

Vol. 144: Seminar on Differential Equations and Dynamical Systems, II. Edited by J. A. Yorke. VIII, 268 pages. 1970. DM 20,–

Vol. 145: E. J. Dubuc, Kan Extensions in Enriched Category Theory. XVI, 173 pages. 1970. DM 16,–

Vol. 146: A. B. Altman and S. Kleiman, Introduction to Grothendieck Duality Theory. II, 192 pages. 1970. DM 18,–

Vol. 147: D. E. Dobbs, Cech Cohomological Dimensions for Commutative Rings. VI, 176 pages. 1970. DM 16,–

Vol. 148: R. Azencott, Espaces de Poisson des Groupes Locale Compacts. IX, 141 pages. 1970. DM 14,–

Vol. 149: R. G. Swan and E. G. Evans, K-Theory of Finite Group Orders. IV, 237 pages. 1970. DM 20,–

Vol. 150: Heyer, Dualität lokalkompakter Gruppen. XIII, 372 S 1970. DM 20,–

Vol. 151: M. Demazure et A. Grothendieck, Schémas en Grou (SGA 3). XV, 562 pages. 1970. DM 24,–

Vol. 152: M. Demazure et A. Grothendieck, Schémas en Group (SGA 3). IX, 654 pages. 1970. DM 24,–

Vol. 153: M. Demazure et A. Grothendieck, Schémas en Group (SGA 3). VIII, 529 pages. 1970. DM 24,–

Vol. 154: A. Lascoux et M. Berger, Variétés Kähleriennes Compa VII, 83 pages. 1970. DM 8,–

Vol. 155: Several Complex Variables I, Maryland 1970. Edite J. Horváth. IV, 214 pages. 1970. DM 18,–

Vol. 156: R. Hartshorne, Ample Subvarieties of Algebraic Vari XIV, 256 pages. 1970. DM 20,–

Vol. 157: T. tom Dieck, K. H. Kamps und D. Puppe, Homotopiethe VI, 265 Seiten. 1970. DM 20,–

Vol. 158: T. G. Ostrom, Finite Translation Planes. IV. 112 pages. 1 DM 10,–

Vol. 159: R. Ansorge und R. Hass. Konvergenz von Differenzen fahren für lineare und nichtlineare Anfangswertaufgaben. VIII, Seiten. 1970. DM 14,–

Vol. 160: L. Sucheston, Constributions to Ergodic Theory and P bility. VII, 277 pages. 1970. DM 20,–

Vol. 161: J. Stasheff, H-Spaces from a Homotopy Point of VI, 95 pages. 1970. DM 10,–

Vol. 162: Harish-Chandra and van Dijk, Harmonic Analysis on Re tive p-adic Groups. IV, 125 pages. 1970. DM 12,–

Vol. 163: P. Deligne, Equations Différentielles à Points Singul Reguliers. III, 133 pages. 1970. DM 12,–

Vol. 164: J. P. Ferrier, Seminaire sur les Algebres Complètes. II, 69 ges. 1970. DM 8,–

Vol. 165: J. M. Cohen, Stable Homotopy. V, 194 pages. 1970. DM 16

Vol. 166: A. J. Silberger, PGL_2 over the p-adics: its Representati Spherical Functions, and Fourier Analysis. VII, 202 pages. 1 DM 18,–

Vol. 167: Lavrentiev, Romanov and Vasiliev, Multidimensional Inv Problems for Differential Equations. V, 59 pages. 1970. DM 10,–

Vol. 168: F. P. Peterson, The Steenrod Algebra and its Applicati A conference to Celebrate N. E. Steenrod's Sixtieth Birthday. 317 pages. 1970. DM 22,–

Vol. 169: M. Raynaud, Anneaux Locaux Henséliens. V, 129 pages. 1 DM 12,–

Vol. 170: Lectures in Modern Analysis and Applications III. Edite C. T. Taam. VI, 213 pages. 1970. DM 18,–

Vol. 171: Set-Valued Mappings, Selections and Topological Propert of 2^X. Edited by W. M. Fleischman. X, 110 pages. 1970. DM 12,–

Vol. 172: Y.-T. Siu and G. Trautmann, Gap-Sheaves and Exten of Coherent Analytic Subsheaves. V, 172 pages. 1971. DM 16,–

Vol. 173: J. N. Mordeson and B. Vinograde, Structure of Arbit Purely Inseparable Extension Fields. IV, 138 pages. 1970. DM 14,

Vol. 174: B. Iversen, Linear Determinants with Applications to Picard Scheme of a Family of Algebraic Curves. VI, 69 pages. 19 DM 8,–

Vol. 175: M. Brelot, On Topologies and Boundaries in Potential The VI, 176 pages. 1971. DM 18,–

Vol. 176: H. Popp, Fundamentalgruppen algebraischer Mannigfal keiten. IV, 154 Seiten. 1970. DM 16,–

Vol. 177: J. Lambek, Torsion Theories, Additive Semantics and Rin of Quotients. VI, 94 pages. 1971. DM 12,–

Vol. 178: Th. Bröcker und T. tom Dieck, Kobordismentheorie. 191 Seiten. 1970. DM 18,–

Vol. 179: Seminaire Bourbaki – vol. 1968/69. Exposés 347-363. 295 pages. 1971. DM 22,–

Vol. 180: Séminaire Bourbaki – vol. 1969/70. Exposés 364-381. 310 pages. 1971. DM 22,–

Vol. 181: F. DeMeyer and E. Ingraham, Separable Algebras o Commutative Rings. V, 157 pages. 1971. DM 16.–